The Book of **Inventions**

The Book of

Inventions

Ian Harrison

NATIONAL GEOGRAPHIC

Washington, DC

One of the world's largest nonprofit scientific and educational organizations, the National Geographic Society was founded in 1888 "for the increase and diffusion of geographic knowledge." Fulfilling this mission, the Society educates and inspires millions every day through its magazines, books, television programs, videos, maps and atlases, research grants, the National Geographic Bee, teacher workshops, and innovative classroom materials. The Society is supported through membership dues, charitable gifts, and income from the sale of its educational products. This support is vital to National Geographic's mission to increase global understanding and promote conservation of our planet through exploration, research, and education.

For more information, please call 1-800-NGS LINE (647-5463) or write to the following address:

National Geographic Society
1145 17th Street N.W.
Washington, D.C. 20036-4688 U.S.A.

Visit the Society's Web site at www.nationalgeographic.com.

Text and design copyright © Cassell Illustrated 2004
First published 2004

Library of Congress Cataloging-In-Publication Data

Harrison, Ian, 1965-
 The book of inventions : how'd they come up with that /
 by Ian Harrison
 p. cm.
Includes bibliographical references and indexes.
 ISBN 0-7922-8296-5
 1. Inventions–History. 2. Inventors–Biography. I. Title.

T15.H3448 2004
609–dc22   2004049922

Printed in China

Contents

Art Fry Inventor of Post-it Notes

FOREWORD

People are full of ideas. An inspiration can come in an instant, but finished, marketed products do not just fall out of the sky. They are far from accidents. They need time and resources. Inventing is about wading through the maze of patents, regulations, market research, investment, manufacturing, and selling, which takes a lot of work and courage. Unexpected problems pop up to block the path and must be solved. On average, 3,000 to 5,000 raw ideas birth only one successful new product. The rest die in various stages of development.

So what motivates inventors to do this work? Necessity is still a big one. Curiosity can start us down a path and we want to see beyond the next turn. Some can't resist unsolved problems and challenges. The chance for a monetary reward, or the discovery of a new technology gets others started. Inventiveness is simply a visceral drive that has its own rewards and satisfactions.

Every species on Earth has been inventive: evolving and adapting to inhabit some particular niche in the world. For most, it required eons for their bodies to adapt, but human inventiveness has found ways to live in all corners of our planet. It's basic to our survival.

Another basic is we all earn a living by producing a needed product or service. Modern machines and tools give individuals enormous productivity. A relatively small number of people can produce the world's supply of a single product. A huge variety of products and services are needed to keep everyone employed. The fruit of our inventiveness is a plenitude of products and services that have given us a standard of living unavailable to the richest kings of old.

What is the relationship between creativity, invention, and innovation? These words get used interchangeably, but they are different. Creativity is a new idea, a flash of inspiration that identifies a new pattern. It could have been there all along, undiscovered, or we can synthesize a new pattern in our minds. Invention is making a prototype of the idea to show others and test to see if it works. Finally, an innovation is an invention that changes old patterns and has been put to use by the customers.

We all learn and recognize patterns. The dog hears the doorbell ring and runs to the door. People learn languages to communicate. We learn crafts and skills for work, and agree on laws and manufacturing standards. Patterns make our lives simpler, predictable, and more efficient. Existing patterns serve us well for most tasks, but sometimes we require or discover a new pattern that works better.

In a choir loft I had a creative idea to make a bookmark with an adhesive that would stick to paper well enough to hold on, but not so tightly as to pull out the paper fibers when it was moved. When I made those bookmarks and discovered their usefulness as sticky notes, they became an invention. When I tested them for every application to be sure they would work and when we had worked out the manufacturing and put them onto store shelves, they were still just an invention. It was only when you learned about Post-it Notes, bought a pad, used it up, and bought a second pad that it became an innovation. You had a new pattern for communicating and organizing. I can't tell what old pattern you gave up because people are always telling me, "I don't know what I did before there were Post-it Notes." They were a pioneer product.

There are patterns to the innovation process. It progresses from pioneer products to line additions to product improvement. Pioneer products are new to the world. They create new markets, jobs, and incomes. They lead to line additions that are complementary products or services that use most of the same technologies and infrastructure, but don't cannibalize the pioneer product. In the case of Post-it Notes, we added Post-it Easel Pad, Post-it Custom Printed Note, Post-it Flag, Post-it Pop-up Note Dispenser, and Post-it Software Note. As a business matures, we innovate with product improvements, cost reductions, or better distribution. This will protect the products from competitors, but seldom grows the market.

An individual, or a small group of people with broad skills, can create the pioneer product with a relatively small capital investment. Mature products need specialists and huge investment in manufacturing and distribution infrastructure. Henry Ford could make every part of his first car by himself, but he wouldn't have been able to repair a modern car.

What do I see as the future of invention? It is precisely what we will need to solve problems presented by population growth, equalized living standards, and dwindling resources. Eighty years ago, no one could predict the microelectronics that would lead to computers and cell phones, or bioengineering or Post-it Notes. The future will present us with both problems and huge opportunities, but inventors will be there for us with solutions that will make us say, "Why didn't I think of that?"

Art Fry

Chapter One

AROUND THE HOUSE

Geography has a great deal to do with who is credited with inventing the light bulb. West of the Atlantic, historians usually cite American inventor Thomas Edison; east of the Atlantic, credit usually goes to English chemist Joseph Swan.

Did you know?
Not only did Thomas Edison fail to secure a British patent after legal action from Sir Joseph Swan, but he lost his American one, in October 1883, when the U.S. Patent Office ruled that a prior invention by William Sawyer and Albon Man (patent granted in June 1878) took precedence.

LIGHT BULB

There is considerable confusion and acrimony over whether it was Sir Joseph Swan (England) or Thomas Edison (U.S.) who invented the light bulb, but in fact their stories are inseparable—and in truth neither of them actually "invented" the light bulb.

Sir Humphrey Davy (England) first demonstrated the principle of incandescent light in 1801: that passing electricity through a filament will cause the filament to "incandesce" or glow brightly. But filaments exposed to the air quickly burned away, so inventors began to search for a way of isolating the filament from the air. During the 19th century some 18 inventors produced experimental incandescent bulbs before 1879, when Swan and Edison made their famous demonstrations.

As early as 1860, Swan patented the first experimental bulb to have a carbon filament. He announced the invention of a practical incandescent light bulb at a meeting of the Newcastle upon Tyne Chemical Society on December 18, 1878, but the filament had blown, and it wasn't until January 18, 1879 that he was able to make the world's first public demonstration, which he did during a lecture in Sunderland, England.

Meanwhile, in America, Edison was also investigating the light bulb, and successfully tested Model No.9 in his laboratory on October 21, 1879, 10 months after Swan's demonstration. Edison filed a patent for his bulb on November 1, 1879 (granted January 27, 1880) and made a public demonstration on New Year's Eve. Swan did not file his patent until November 27, 1880, more than a year after Edison's application in the U.S.

Edison produced the world's first commercially available light bulbs in October 1880, a few months before Swan began production in England. When Edison applied for a British patent Swan began legal proceedings for patent infringement, but they settled out of court and, in 1883, joined forces to form the Edison & Swan United Electric Company, whose bulbs were marketed under the name "Ediswan."

See also: Hairdryer, page 40; Phonograph/Kinetoscope, page 158

Left X-ray of an electric light bulb **Above** Students at Cleckheaton Grammar School, Yorkshire, England, use a light bulb in an electrical experiment (May 1956)

The co-inventor: Joseph Wilson Swan

1828 Born on October 31, near Sunderland, England

1841 Apprenticed to a druggist

1846 Joins a pharmaceutical company in Newcastle upon Tyne and later becomes a partner in the company

1856 Patents an improved wet-plate collodion photographic process

1857 Patents high-speed bromide photographic paper, an idea that he sells to George Eastman, founder of Kodak

1860 Patents the first incandescent light bulb to have a carbon filament

1874 Elected a Fellow of The Royal Society

1879 Makes the world's first demonstration of a practical incandescent light bulb

1880 Patents an improved version of his incandescent light bulb

1883 Forms Edison & Swan United Electric Company with Thomas Edison. Patents an improved filament using nitrocellulose thread. This is the world's first synthetic fiber

1904 Receives a knighthood. Patents improvements to intaglio printing methods

1914 Dies on May 27, in Warlingham, Surrey, aged 85

For biography of Thomas Alva Edison see page 158

Did you know?

When electric light was first introduced, people were so used to gas lighting that the Edison Company issued signs carrying instructions and reassurance, reading: "This room is equipped with Edison Electric Light. Do not attempt to light with match. Simply turn key on wall by the door. The use of electricity for lighting is in no way harmful to health, nor does it affect soundness of sleep."

The electric iron was invented by New Yorker Henry W. Seely and patented on June 6, 1882. Seely's invention was somewhat ahead of its time, since America's first power station did not begin generating electricity until three months later.

ELECTRIC IRON

The idea of removing the creases from clothes is centuries old. It began with smoothing tools made of wood, glass, or marble and the process improved with the addition of heat: Heavy stones would be warmed on the hearth and used to press the clothes. The earliest mention of a laundry iron dates from the 17th century—a smooth block of iron with a handle was a much better tool than a heavy stone. The first irons were heated next to an open fire, but later developments included hollow irons, which could be filled with embers or charcoal to keep them hot, and by the 19th century it was standard practice to heat the iron on a stovetop.

Henry W. Seely (U.S.) realized that there was another way to heat the footplate of an iron, by using an electric current. In 1882 he invented the electric iron, which was heated by an arc of electricity bridging the gap between two carbon rods set in the hollow base of the iron. Seely improved on this in 1883 with a much safer, cordless electric iron. This worked on the same principle as modern cordless irons—it was plugged into a special stand to heat up and was then removed from the stand to enable the ironing to be done without the encumbrance of a cable. However, it was expensive and unreliable, which, coupled with the fact that very few people had an electricity supply, meant that it was not a commercial success.

Electric irons gained acceptance as they gradually improved, and then in 1926 New York dry cleaning company Eldec (U.S.) introduced the first domestic steam iron. Like Seely's pioneering iron before it, the Eldec iron was ahead of its time. Technology had not reached the point where the steam iron could be made safe enough or reliable enough to be an immediate success, but the idea has since proved to be sound, and by the end of the century more than 80 percent of electric irons sold were steam irons.

Did you know?

Today's irons have Teflon-coated nonstick footplates, but in Elizabethan times beeswax and candle grease were often used to help the iron slide over linen and other fabrics.

In 2001 Corpo Nove (Italy) invented a wrinkle-free shirt, known as the Oricalco, made of titanium-alloy fibers interwoven with nylon. The fabric reshapes when heated, and simply blowing the wrinkled shirt with a hairdryer dissolves any creases.

Trousers with a permanent crease were invented by CSIRO in Australia and first produced in 1957, using a process that was patented under the name Si-ro-set.

Electric iron time line

1882 Henry W. Seely (U.S.) invents the electric iron, heated by a carbon arc. (A carbon arc iron is invented independently in France shortly afterward.) One of the problems with carbon arc irons was that pieces of red-hot carbon often showered the clothes, burning myriad tiny holes

1883 Seely invents a safer, cordless electric iron

1891 GEC and Crompton (both England) sell electric irons with flexible cords to connect them to a light socket

1926 Eldec Co. (U.S.) introduces the first domestic electric steam iron

1936 The first successful thermostatically controlled electric iron goes on sale

Opposite top "Clem" universal traveling iron (c. 1954) **Below** Dancing for joy over the new electric iron

James Dyson is famous as the inventor of the Dual Cyclone vacuum cleaner and, more recently, of the Contrarotator washing machine, but "inventor" is not his favorite word—he prefers to be called an engineer or a product designer.

The inventor: James Dyson

1947 Born in Norfolk, England, the youngest of three children, and later trains as a furniture designer at the Royal College of Art, London

1970 Wins a Design Council award for his goods-carrying boat, the Sea Truck

1975 Launches the Ballbarrow, a variation on the traditional wheelbarrow. In his *History of Great Inventions*, Dyson writes: "It was only after spending days using a wheelbarrow while renovating my house in Bath that I became aware that the

DUAL CYCLONE CLEANER

James Dyson was born in Norfolk, England, into what he describes as a "middle-class and not particularly wealthy" family. He trained as a furniture designer at the Royal College of Art in London before taking his designer's aesthetic with him into the field of engineering. One of his earliest engineering projects was the Sea Truck, a goods-carrying boat that in 1970 won him an award from the Design Council. Five years later he launched the Ballbarrow, a variation on the wheelbarrow—as the name suggests, it had a large ball instead of the wheel, making it more stable than a conventional wheelbarrow and preventing it from sinking into soft ground or damaging lawns.

Production of the Ballbarrow led indirectly to Dyson's best known invention, the Dual Cyclone vacuum cleaner. The air filters in the spray-finishing room of the Ballbarrow factory were always clogging up, so the factory installed a cyclone-generating device to remove the particles from the air by centrifugal force. This in itself was not a new idea, but Dyson began to ponder the possibilities of applying the same principle to a domestic vacuum cleaner, having become frustrated by the rapid loss of suction that occurs as the bags of conventional vacuum cleaners fill with dust.

Dyson spent 15 years and built 5,127 prototypes in developing the Dual Cyclone vacuum cleaner, which was so named because two cyclones provided the suction (unlike conventional vacuum cleaners in which a fan sucks air through a bag). Dust and dirt are flung out of the airstream by the centrifugal force into a container, which ensures that the suction does not diminish as the cyclonic cleaner fills up. As with the Ballbarrow, Dyson had identified the least efficient part of the conventional machine and dispensed with it. Or, in his own words: "I became determined to come up with a vacuum cleaner with a performance that did not stand or fall by the properties of a paper bag."

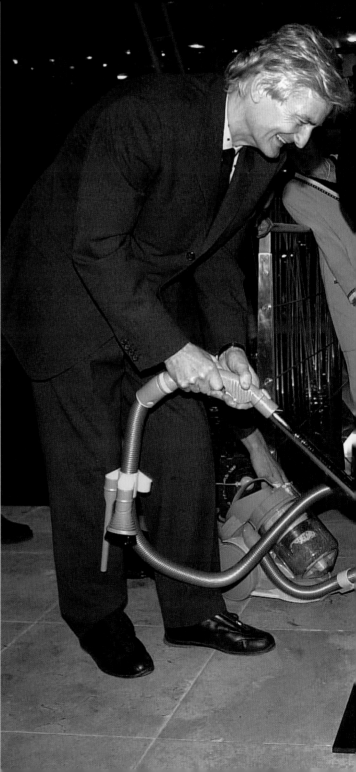

2,000-year-old design could use some improvement. And, what's more, improvement of the component that is widely regarded as the greatest of all inventions—the wheel."

1975–78 Produces the Trolleyball, a trolley with balls instead of wheels, and an amphibious vehicle known as the Wheelboat

1978–93 Develops the Dual Cyclone vacuum cleaner (patent filed 1979, granted 1983)

2001 Launches the Contrarotator washing machine, the world's first two-drum washing machine. Introduces the Root Cyclone vacuum cleaner, which uses a number of cyclones to produce greater and more constant suction than even the Dual Cyclone

2003 Exhibits an uphill waterfall at the Chelsea Flower Show, London

Left Queen Elizabeth II and Prince Phillip look on as James Dyson vacuums a plaque unveiled to mark the Royal visit to his factory (December 2001) **Above** James Dyson in London (2002)

Did you know?

Dyson's hero is engineer Isambard Kingdom Brunel. His ambition, as stated in his autobiography, is that: "One day, Dyson will replace 'Hoover' and become a noun, a verb, out there on its own, long after I am forgotten."

Dyson products have been exhibited at the Victoria & Albert Museum (V&A) in London; the San Francisco Museum of Modern Art; the Georges Pompidou Centre in Paris; and the Powerhouse Museum in Sydney, Australia. His products are also on display at the Design Museum in London, of which Dyson was appointed chairman in 1999.

Previous owners of James Dyson's converted mill house in Wiltshire include rock star Van Morrison and film producer David Puttnam.

Andrew Ure's 1830 patent relating to: "An apparatus for regulating temperature in vaporisation, distillation and other processes" not only marked the invention of such a device, but also contained the first recorded use of the word "thermostat."

THERMOSTAT

Thermostat may have been a new word and a new concept in 1830, but the technology that went into this temperature-regulating device was not. Andrew Ure (Scotland), inventor of the thermostat, describes several types of devices in his patent, the most sophisticated involving a bimetallic strip, which is still the basis of most modern thermostats (often in the form of a bimetallic disk). The idea of a bimetallic strip is that as the temperature changes the two conjoined metals expand or contract to different degrees, causing the strip to bend, activating or deactivating a controller. Ure may have been the first to apply the principle to temperature control, but the pioneer of such devices was John Harrison (England), who had used a bimetallic pendulum in his famous chronometer of 1726, to compensate for inaccuracies caused by changes in temperature.

All the ideas in Ure's patent remained obscure, and the first thermostat to come into general use was a liquid thermostat patented by Charles Edward Hearson (England) in 1881. Hearson's thermostat was designed for use in poultry incubators, and consisted of a closed container of a liquid that boiled at the required temperature, causing the container to expand and activate a control lever. The same principle remains in use today, although the simple expansion or contraction of the liquid in a capillary tube is sufficient (and more versatile) than requiring liquids that boil at a particular temperature as Hearson envisaged.

A third type of thermostat uses a bellows filled with liquid or gas, an invention patented by W.M. Fulton (Britain) in 1903. The bellows expands or contracts as the temperature of the liquid or gas filling it rises or falls, which has the added advantage that it can be used to provide a gradual temperature change (e.g., by controlling the flow of gas to an oven).

Some of the places to find a thermostat are in central heating systems, electric irons, dishwashers, washing machines, kettles, fridges, freezers, and ovens.

See also: Incubator, page 220

Above Thanks Andrew Ure for warm rooms in winter and cool homes in summer
Opposite The electro-magnetic regulator, an early thermostat (c. 1870)

The inventor: Andrew Ure

1778 Born in Glasgow, Scotland, on May 18, the son of a cheesemonger

1801 Graduates from the University of Glasgow

1804 Becomes Professor of Natural Philosophy at the Andersonian Institution (now the University of Strathclyde)

1821 Publishes *Dictionary of Chemistry*

1822 Elected a Fellow of The Royal Society

1829 Publishes *System of Geology*

1830 Patents the first thermostat

1835 Publishes *Philosophy of Manufactures*

1839 Publishes *Dictionary of Arts, Manufactures and Mines*

1857 Dies in London, England, on January 2, aged 78

DISHWASHER

More than 30 American women patented dish-washing machines in the last three decades of the 19th century. The most successful, and the first to be manufactured commercially, was invented by Josephine Cochran, who filed her patent in 1885.

Josephine Cochran (U.S.) spelled her name in various ways (sometimes "Cochrane," sometimes "Cockran"), which is said to be because she aspired to greater social status and was looking for a more refined version of her surname. But whatever the variations she used elsewhere in her life, the name that appears on the patent application for her "dish washing machine" is Josephine G. Cochran.

The idea of a dish-washing machine was clearly a preoccupation for American women in the late 19th century, and particularly for Josephine Cochran. Her social aspirations meant that she considered washing her own crockery beneath her and she did not trust her servants to treat her best dinner service with the care it deserved, so she set about inventing a machine to do the washing up for her. Perhaps the fact that she was descended from John Fitch (who a century earlier had made a disputed claim to have invented steam navigation) gave her the inventive edge over her competitors; on December 31, 1885, after six years of development, she filed a patent for what was to become the first commercially manufactured dishwasher.

Hot water and soap suds were pumped out of two separate cylinders in the bottom of the machine, alternately dousing the racks of crockery before draining back to the cylinders to be pumped round again. On domestic models the piston-pumps were driven by a handle at the side of the machine, while the larger models, intended for hotels and restaurants, were steam-powered. After her machines were exhibited at the 1892–93 Columbian Exhibition in Chicago, one newspaper reported that the machines were "capable of washing, scalding, rinsing, and drying from 5 to 20 dozen dishes of all shapes and sizes in two minutes."

Having achieved success, Cochran said that trying to market her machine had been harder than actually inventing it, and announced: "If I knew all I know today I would never have had the courage to start."

Did you know?

One early dishwasher, invented by Eugène Dauguin (France) for a restaurant in Paris, comprised eight artificial hands that bobbed the dishes and plates up and down in the water to clean them.

Cochran's Garis-Cochran Dish-Washing Machine Company eventually became part of the Whirlpool Corporation.

An anonymous 19th-century epitaph in an English churchyard read:
Here lies a poor woman who was always tired,
For she lived in a place where help wasn't hired.
Her last words on Earth were, Dear friends I am going
Where washing ain't done nor sweeping nor sewing,
And everything there is exact to my wishes,
For there they don't eat and there's no washing of dishes...
Don't mourn me now, don't mourn me never,
For I'm going to do nothing for ever and ever.

The inventor: Josephine Cochran

1841 Born Josephine Garis in Ohio, the daughter of a civil engineer (birthdate sometimes given as 1839). Later marries merchant and politician William Cochran

1879 Begins developing her dishwasher

1885 Files a patent on December 31 for her dishwasher (U.S. patent granted December 28, 1886, U.K. patent granted 1887)

1888 Patents improvements to her dishwasher

1889 Sells manufacturing rights to Crescent Washing Machine Company (U.S.)

1892–93 Exhibits her dishwashers at the Columbian Exhibition in Chicago, U.S.

1907 Patents improvements to her dishwasher

1913 Dies

Opposite top Housewives once dreaded washing dishes by hand **Below** The view from inside a modern dishwasher

Domestic food processors evolved from simple food mixers to multipurpose processors. There is no single inventor of the food processor, but the first electric mixers were made by Hobart Manufacturing and the Hamilton Beach Manufacturing Co.

Food mixers/processors time line

1908 Herbert Johnson (U.S.) designs a baker's mixer for Hobart Manufacturing (U.S.)

1910 Chester A. Beach, Fred Osius, and L.H. Hamilton of Hamilton Beach Manufacturing Co. (all U.S.) produce the first patented food mixer

1916 Hobart Manufacturing begins supplying the U.S. Navy, and by 1917 the mixer is classified as standard equipment on U.S. Navy ships

1919 Troy Metal Products begins producing the H-5 Mixer

FOOD PROCESSOR

Chester A. Beach (U.S.) was a pioneer of small, high-speed, low-power universal electric motors and made his first "fractional horsepower" motor in 1905. The diminishing size of electric motors enabled them to be used in a number of new appliances, and in 1910 Beach formed the Hamilton Beach Manufacturing Co. with Fred Osius and L.H. Hamilton (both U.S.) in order to market the first patented electric food mixer. However, two years earlier Herbert Johnson (U.S.) of Hobart Manufacturing had designed a baker's mixer to take the labor out of kneading dough, and it was this earlier mixer that led the way toward multipurpose food processors.

Hobart Manufacturing began supplying the U.S. Navy in 1916. The mixer proved so popular that in 1919 the company began producing a domestic version, the Model H-5 Mixer, through a subsidiary called Troy Metal Products. The H-5 was the first domestic mixer with its own fixed stand and bowl, and the first in which the beater and the bowl rotated in opposite directions, a feature known as "planetary action" (a patent for this feature was granted in 1920). The H-5 was soon renamed KitchenAid, a brand still going strong today. During the 1930s Egmont Arens (U.S.) designed a smaller, sleeker, lighter version, now acknowledged as a design classic.

In 1922 Stephen J. Poplawski (U.S.) invented the first "blender" by putting a spinning blade in the bottom of the bowl rather than having beaters/mixers coming into the bowl from above. The next big step came in 1948, when Ken Wood (England) produced a mixer that by 1950 he had developed into the first multipurpose food processor —a combined mixer, siever, blender, juicer, grinder, liquidizer, mincer, peeler, slicer, and shredder known as the Kenwood Chef. The first of the modern, ultracompact, lightweight domestic food processors was the Magimix, invented by Pierre Verdun (France) in 1971. It was not a huge success in France but became very popular in America, where it was known as the Cuisinart.

See also: Hairdryer, page 40

1920 Patent granted for the H-5's "planetary" mixing action

1922 Stephen J. Poplawski (U.S.) invents the first blender

1950 Ken Wood (England) invents the first multipurpose food processor

1971 Pierre Verdun (France) produces the first modern, ultracompact, lightweight, domestic food processor—the Magimix

Did you know?

The world's largest bowl of soup contained 662 gal (2,505.94 l) of beef and vegetable soup, and was made by Wyler's (U.S.) in 2000 to celebrate the company's 70th anniversary. The largest guacamole was made in Michoacan, Mexico, on November 26, 2000, using 2,205 lb (1,000 kg) of avocados, 243 lb (110 kg) of onions, 243 lb (110 kg) of tomatoes, 33 lb (15 kg) of chilies, 11 lb (5 kg) of coriander, 26 lb (12 kg) of salt, and 79 lb (36 kg) of limes.

Legend has it that the name KitchenAid was coined when wives of the Troy Metal Products executives were testing the H-5 Mixer. One of them said: "I don't care what you call it, it's the best kitchen aid I have ever had."

Racine, Wisconsin, home of the Hamilton Beach Manufacturing Co., was also the home of the Horlick Malted Milk Co., and it was at Horlick's suggestion that Beach developed the famous Hamilton Beach drink mixer in 1911.

Right 1950s advertisement for the Kenwood Chef, stating: "More than a mixer at Christmas *and* all through the year. The Kenwood Chef—Your Servant Madam!"
Left A sleek modern electric blender

The microwave oven was invented by Percy LeBaron Spencer, who was one of a number of wartime radar scientists to notice the heating effects of microwaves. Others dismissed the phenomenon, but Spencer used it to revolutionize cooking practices.

MICROWAVE OVEN

Percy LeBaron Spencer (U.S.) is said to have discovered the culinary potential of microwaves after finding a melted chocolate and peanut bar in his pocket— he had just walked past a microwave-emitting device called a magnetron, and he decided to investigate whether that is what had melted the chocolate bar. The cavity magnetron had been developed in the early 1940s by physicists Sir John Randall and Dr. H.A.H. Boot (both England) as a means of producing microwaves for use in airborne radar, and their invention was patented in 1947. This new radar technology was shared with wartime ally the U.S., where President Roosevelt, well aware of the tactical importance of the cavity magnetron, described it as "the most valuable cargo ever to reach our shores."

Spencer, working for the Raytheon Manufacturing Co. in Newton, Massachusetts, suggested various improvements to the device, and Raytheon duly won the contract to manufacture magnetrons. Noticing the heating effect of the microwave beam, which is now known to be due to the excitation of molecules rather than the radiation of heat, Spencer investigated further. He put a bag of maize in the path of the microwave beam, and seconds later he had a bag of popcorn. Then he created the world's first microwave oven by cutting a hole in the side of a kettle and directing the microwave beam from the magnetron through the hole. An egg placed in the kettle cooked so quickly that it exploded in spectacular fashion, and this demonstration was enough to persuade Raytheon to begin work on developing Spencer's invention.

On October 8, 1945, Spencer filed a patent for a "method of treating foodstuffs" (granted 1950), and the first prototype microwave oven was installed in a restaurant in Boston the following year. The prototype was a success, and Raytheon produced its first commercial microwave ovens in 1947, known as the Raydarange.

Above A hamburger cooked in one of the first Raydarange microwave ovens (1947) **Opposite** Inside a microwave oven

The inventor: Percy LeBaron Spencer

1894 Born on July 19 in Howland, Maine

1910 As part of a four-man team, Spencer installs an electrical system in a local paper mill, largely by trial and error

1912 Joins U.S. Navy to learn wireless telegraphy

1920s–40s Works for Raytheon Manufacturing Co. (U.S.), during which time he patents a High Efficiency Magnetron

1945 Files a patent for the microwave oven (granted 1950)

1970 Dies on September 8, aged 76

See also: Radar, page 110

Did you know?

The first microwave ovens weighed about 772 pounds (350 kilograms), were 6 feet (2 meters) tall, and cost nearly $5,000 each.

Popcorn is held in great affection in America. Percy Spencer found a place for it in the history of the microwave oven, and in 1992 food writer Betty Fussell (U.S.) commented that: "Popcorn is a truly indigenous fast finger-food that links all ages, places, races, classes, and kinds in the continuing circus of American life.

Popcorn is the great equalizer, which turns itself inside out to attest to our faith that color is only skin deep and class superfluous."

Like many inventions, the refrigerator developed from earlier ideas. But if anyone can be said to have invented the machinery that led to the modern fridge, it is Jacob Perkins, who, in 1834, patented "apparatus for producing ice and cooling fluids."

REFRIGERATOR

Nature provided the first methods of refrigeration, most obviously through the transfer of heat to melting snow or ice, but also through the process of evaporation. The human body is cooled by evaporation, which is why we sweat, and the ancient Greeks discovered that food kept in damp earthenware jars could be kept cool by the same process, a method still in use today in some parts of the world. As early as 1775, chemist William Cullen (Scotland) recorded experiments that took the natural process of evaporation a step further, obtaining freezing temperatures by evaporating nitrous ether under reduced pressure.

More than half a century later, mechanical engineer and inventor Jacob Perkins (U.S.) was the first person to build a machine that put Cullen's principle to practical use. Perkins not only evaporated nitrous and sulfuric ether under reduced pressure, he also compressed the resulting gas in order to reliquefy it so that it could be used to continue the cooling process by evaporating again; he had thus put together the essential components of a modern compression fridge. Because he was living and working in London, in 1834 Perkins took out a British patent for his invention rather than an American one.

Perkins did not exploit the commercial potential of his invention, but its principles led to the development of industrial refrigeration simultaneously in America, by Alexander Twining (U.S.), and in Australia, by James Harrison (Scotland–Australia). Commercially operated refrigeration plants opened in both countries in 1850, but the first domestic refrigerator did not go on sale until 1913. This was the "Domelre" (_dom_estic _el_ectric _re_frigerator), which was manufactured and first sold in Chicago—the drawback from an aesthetic point of view was that the refrigeration unit was mounted on top, making it look more like a piece of industrial machinery than a desirable domestic appliance.

Left Advertisement for White Mountain Refrigerators (1903)
Above Jacob Perkins (c. 1810), inventor of the earliest refrigerator

The inventor: Jacob Perkins

1766 Born on July 9 in Newburyport, Massachusetts

1769 Apprenticed to a goldsmith, under whose auspices he makes dies (i.e., metal tools or stamps) for the state coinage

1780s Develops tempered steel engraving plates for printing banknotes and, later, the first transfer printing press

1818 Moves to England, where he establishes the engraving factory Perkins, Bacon & Co. and begins experimenting with refrigeration and high-pressure steam boilers

1825 Demonstrates the world's first steam-powered gun

(invented 1824) to the Duke of Wellington in Regent's Park, London, England. The Duke of Sussex describes the gun as "Wonderful, damned wonderful," but a steam boiler is considered impractical for use on a battlefield *See also: Guns, page 254; Maxim gun, page 256*

1831 Invents a water tube boiler

1834 Patents the first compression refrigerator

1840 Perkins, Bacon & Co. (England) prints the world's first adhesive postage stamps

1849 Perkins dies on July 30 in London, England, aged 83

One of many inventions to have been invented in several places at almost the same time, the sprung mousetrap was patented in America by William Hooker in 1894 and in Britain by James Henry Atkinson in 1898.

MOUSETRAP

Mousetraps have been taxing inventors' ingenuity for centuries, and since the mid-19th century there have been more than 4,000 mousetrap patents granted in America alone. The classic is the sprung trap, or "snap trap," which has been immortalized by cartoons and is still in everyday use more than a century after its invention. The first sprung trap was invented by William Hooker (U.S.), patented in 1894, and marketed as the "Out O' Sight," with a trademark featuring a mouse peering out of the middle "O." (Hooker patented 27 inventions between 1865 and 1908; the first was for a hedge trimmer, but all the rest related to gates or animal traps.) Five years later, John Mast (U.S.) invented a very similar device and set up a factory in Lititz, Pennsylvania (now Woodstream Corp.), to market the Victor, which today is said to be the world's best-selling mousetrap.

Four years after Hooker patented the Out O' Sight (and a year before Mast's Victor), James Henry Atkinson (England) patented another sprung mousetrap. Whether he invented his trap independently or copied Hooker's is open to debate, but it is almost certain that under new rules introduced by the British Patent Office in 1905, he would not have been able to patent his invention because it was not novel. Atkinson had previously filed patents relating to window blinds, fireplaces, and laundry irons. Then, on December 30, 1898, he filed a patent for an: "Improved treadle trap for mice, rats and the like." This patent mitigates the suggestion that he copied his trap from Hooker, because the treadle trap relied on mice simply running across the trap—it was Atkinson's 1899 patent for an improved version that described the classic "Little Nipper" mousetrap, which, according to the patent, is "caused to 'go off' by a pull at the bait." The Little Nipper was, and still is, made by the Welsh company Procter Brothers at its wire-working factory in Gwent.

"Little Nipper" inventor: James Henry Atkinson

1849 Born in Leeds, England

1886 Patents a rack pulley for window blinds; his patent describes him as an ironmonger

1893 Patents a cinder shifter for fireplaces

1895 Files a patent (later abandoned) for window blind cords

1897 Files a patent (later abandoned) for heating laundry irons

1898 Files two patents for mousetraps, the first abandoned and the second granted

1899 Files a patent on June 27 for what becomes the "Little Nipper"; his patent now describes him as a model-maker

1900–1938 Patents several improvements and variations on rodent traps

1942 Dies in Leeds, aged 93

Left Cheese, please **Above** Only a few lucky mice can outsmart Victor mousetraps

Did you know?
American poet and essayist Ralph Waldo Emerson is quoted as saying during a lecture: "If a man write a better book, preach a better sermon, or make a better mousetrap than his neighbor, tho' he build his house in the woods, the world will make a beaten path to his door."

Aerosol cans were seen as a huge boon for convenience until scientists realized that the CFCs used in them as propellants were destroying the ozone layer. Since then, Erik Rotheim's invention has provided continued convenience without the use of CFCs.

AEROSOL CAN

The principle behind the aerosol can is the same as that employed by winning racing drivers—pressurize the gas in a container (e.g., by shaking a magnum of champagne) and any liquid that shares the container will be propelled into the air. Aerosol cans, however, are descended not from champagne bottles but from the metal or glass phials used by 19th-century physicians to dispense fine sprays of antiseptic; the phials would be warmed in the hand to increase the pressure, thereby producing the spray of liquid. The first patent for an aerosol spray was granted to J.D. Lynde (U.S.) in 1862, but, although it was the first, even Lynde did not consider it to be entirely novel: His patent relates to an: "Improved bottle for aerated liquids."

Aerosol cans evolved with patents for improved valves in 1901 (G.L. Bebaur, U.S.), and the use of carbon dioxide as a propellant in 1903 (R.W. Moore, U.S.), but the precursor of the modern aerosol can was invented by Erik Rotheim (Norway) in 1926. Rotheim patented a pressurized container with a special valve, intended for dispensing liquid soap, paint, insecticide, and cosmetics. Commercial production began in Norway soon afterward, but the idea did not catch on, and Rotheim's invention was abandoned.

Then, during the Second World War, the aerosol can was reborn. American troops fighting in the Pacific were suffering more deaths from insectborne diseases than from enemy action, and the Department of Agriculture set about finding a solution. In 1943 Lyle D. Goodhue and William N. Sullivan (both U.S.) patented a "Method of applying parasiticides"; this was an aerosol insecticide better known as the bug bomb. After the war an improved lightweight version was manufactured by Airosol Inc. (U.S.), and went on sale to civilians in November 1946. During the 1970s scientists realized that CFC propellants were damaging the ozone layer; since then CFCs have been phased out—most modern aerosol cans use carbon dioxide, hydrocarbons, or nitrous oxide as propellants.

ow Aerosol perfume by Yardley of Bond Street, London **Above** A portion of the graffiti-covered (courtesy of aerosol cans) Berlin Wall after the wall opened in 1989

Did you know?

Aerosol is the name given to any substance containing tiny particles of liquids or solids suspended in a gas. Clouds and fog are examples of naturally occurring aerosols, in which water droplets are suspended in atmospheric air. Aerosol cans are so called because they produce aerosols when sprayed, so to shorten the term "aerosol can" to "aerosol" is, technically speaking, incorrect.

The inventor: Erik Rotheim

1898 Born Erik Andreas Rotheim in Oslo, Norway

1916 Begins studying at the University of Oslo

1917 Moves to Zürich, Switzerland

1921 Graduates as an engineer at the Swiss Polytechnic School in Zürich

1921–25 Continues his studies in Munich, Germany

1925 Returns to Norway

1926 Files a patent for a pressurized container, the forerunner of the modern aerosol can (Norwegian patent granted June 24, 1927)

1931 Receives his first U.S. patent for "a method of, and agents for, atomizing or distributing liquid and semi-liquid materials"

1932 Receives another Norwegian patent relating to the use of hydrocarbon gases and halogenated hydrocarbons as propellants (halogens are the group of chemicals that includes chlorine and fluorine)

1936 Patents further improvements to his invention

1938 Dies in Oslo, aged 40

1998 On the centenary of Rotheim's birth, the Norwegian Post Office issues a stamp commemorating Rotheim and the invention of the aerosol can

In 1902 an English gunsmith invented a machine to wake someone up with a cup of tea. Then in 1976, perhaps thinking that a cup of tea was too gentle for hardy Canadians, Peter MacNeil patented his "Cold Air Blast Wake-Up Apparatus."

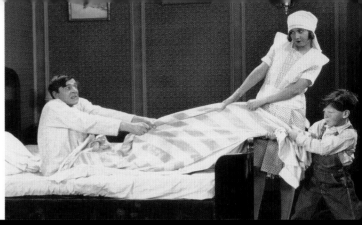

WAKE-UP DEVICES

The purpose of a patent is to prevent other people from copying one's idea. In which case, patenting something called Cold Air Blast Wake-Up Apparatus is a somewhat pointless exercise given that it is a product no one would want to buy and an idea no one would want to copy. As with individuals in other walks of life, many inventors are full of hot air, but on May 24, 1976, Peter MacNeil (Canada) reversed the trend with his invention. It is a fairly self-explanatory variation on the humble alarm clock, but, according to MacNeil's patent: "Such devices as alarm clocks and clock radios have not proved adequately effective to awaken certain individuals having a low sensitivity to aural stimulus when sleeping."

He had seen a niche in the market, and he attempted to fill it. His patent specifies: "Wake-up apparatus includes an elongated hose at one end for coupling to the output side of a room air conditioner and a horizontally elongated rectangular outlet on the other end for placement on a mattress for directing air therealong. A rotatable damper vane mounted on the hose is controlled by a motor energized via an electric timer for driving the vane to an open position at a predetermined time for causing a blast of cold air along the mattress."

Far more amenable was the good old automated teamaker, first invented by Samuel Rowbottom (England), who filed a patent on December 17, 1891 (granted 1892). It was a complex contraption of dubious safety, including as it did a precarious tipping device to decant the boiling water, and it did not catch on. In 1902 gunsmith Frank Clarke (England) invented the first practicable teamaker, although it was still relatively unsafe, having a mechanism involving matches and methylated spirits and retaining the idea of a tipping device. Perhaps not surprisingly, this did not catch on either, and automated teamakers did not become a commercial success until the British Vacuum Cleaner and Engineering Company introduced the Goblin Teasmade in 1937.

See also: Tea bag/Instant coffee, page 54

Did you know?

The earliest form of mechanical wake-up apparatus was the alarm clock—the first known example was made in Würzburg, Germany, c. 1350–80.

Above Silent movie era wake-up call
Right Automated tea-making machine patented by Frank Clarke and made by the Automatic Water Boiler Company (1902)

Tea-maker inventor: Frank Clarke

1876 Born in Birmingham, England, the son of a gunsmith and the sixth of six children

1902 Files a patent for: "An Apparatus Whereby a Cup of Tea or Coffee is Automatically Made." His invention is manufactured by the

Automatic Water Boiler Co. and marketed as "A Clock That Makes Tea"

1902–1930s Runs his own gun manufacturing business in Birmingham

1990s Birmingham history lecturer Chris Upton (England) writes: "I hereby nominate Frank Clarke (posthumously) for the Nobel Peace Prize: the gunmaker who encouraged people to stay in bed and make tea, not war"

The disposable diaper was invented in 1951 by Marion Donovan, a prolific inventor who also patented more than a dozen other inventions to make life easier around the house, including "DentaLoops," the "Big Hangup," and the "Zippity-Do."

DISPOSABLE DIAPER

Marion Donovan (U.S.) was born into an inventive family. Her father, Miles O'Brien, and his twin brother, John, had invented an industrial lathe and founded the South Bend Lathe Co. in 1906, and after her mother died when Marion was seven, she spent much of her time in their factory. She graduated from Rosemont College, Philadelphia, in 1939 with a B.A. in English literature and then worked for *Harper's Bazaar* and as assistant beauty editor for *Vogue* before marrying leather importer James F. Donovan and giving up her job to start a family.

Giving up her job proved to be a good career move, because having children inspired her first invention. Traditional cloth diapers were usually worn with rubber pants to prevent them leaking, but the rubber would trap too much moisture, causing diaper rash. Donovan decided there must be a better solution and, starting with a shower curtain, she experimented with various fabrics before settling on parachute nylon as the best answer; her leakproof diaper cover went on sale in 1949. Meanwhile, she had improved on her own idea by inventing a disposable paper diaper for use with the leak-proof outer cover. The idea was such a success that Donovan could not meet demand and sold the rights to her diaper patents for one million dollars to Keko Corporation (U.S.), which was able to manufacture the diapers in bulk. Donovan's invention of biodegradable paper disposables with nondisposable leakproof covers was later superseded by all-in-one disposable diapers (since criticized for not being biodegradable), which were test-marketed during the 1950s and launched in 1961 by Procter & Gamble, under the name Pampers.

In 1958 Donovan graduated in architecture from Yale University and later designed her own house in Connecticut. She died in 1998, having filed more than a dozen patents for inventions including the Big Hangup (a multigarment compact hanger) and the Zippity-Do (a cord to facilitate fastening dresses that zip up the back).

The inventor: Marion Donovan

1917 Born Marion O'Brien on October 15 in Fort Wayne, Indiana. Later, as a schoolgirl, invents a tooth powder

1939 Graduates from Rosemont College, Philadelphia, with a B.A. in English literature

1946 Invents a reusable leak-proof diaper cover (patent granted 1951)

c. 1949 Invents the first disposable paper diaper

1951 Sells the rights to her diaper patents to Keko Corporation (U.S., children's

Did you know?

Marion Donovan's son James took his mother's interest in diapers several stages further: James F. Donovan, Jr., M.D., is now a urologist.

Marion Donovan once said: "I never thought of myself as an inventor. My invention happened by gradual steps with necessity nudging me." After Marion's death in 1998, her daughter Christine, now a sculptor, said: "I was one of the necessities."

Actress Jamie Lee Curtis patented a diaper featuring a pocket to hold baby wipes.

Opposite top Marion Donovan with husband James and daughters Sharon (7 months) and Christine (2 ¾ years), photographed in 1947
Below Thanks to Marion Donovan, there will be no trace of this escape

clothing manufacturer) for one million dollars

1950s Invents improved facial tissues, a "dispensing container for flowable material," and a nursing bottle marker

1958 Graduates from Yale University School of

Architecture with a master's degree

1960s Invents a tea bag envelope, two-way envelope, shoe box/storage container, improved index cards, the Big Hangup, and the Ledger Check (a combined checkbook and recordkeeping book)

1970s Invents a towel holder, an umbrella, the Zippity-Do and the Turtling-a-ling (a brass turtle with bells on its head and tail, which can be used to simulate a doorbell in order to let phone callers know that other priorities demand attention) *See also: Zipper/Velcro, page 36*

1985 Invents DentaLoops (dental tape formed into loops for easier flossing)

1998 Dies on November 4 in New York, aged 81

In 1849 Walter Hunt was granted a patent for something he called a "dress pin"—a gloriously simple invention better known today as the safety pin. After Hunt's death one wit joked that "without him we would be undone."

SAFETY PIN

It seems that the safety pin is a much older invention than the 1849 patent granted to Walter Hunt (U.S.) would suggest. There is evidence that the Romans used self-sprung safety pins more than two millennia ago, but, like many other Roman inventions, the idea was subsequently forgotten until it was reinvented in modern times. A more recent precursor of Hunt's safety pin was the "Victorian shielded shawl and diaper pin," patented by Thomas Woodward (U.S.) in 1842. This comprised a straight pin hinged to a separate piece of metal that was bent over at the end to cover the point of the pin; very similar to the successful safety pin, but more difficult to use and less reliable.

Hunt's "dress pin" was much simpler. Some inventions come about by accident, others from a moment of inspiration, yet others from years of hard work and development, but the safety pin was the result of a challenge arising from a bad debt. Hunt was a prolific inventor, but he was in debt to the tune of $15 to the draftsmen who prepared his many patent drawings. The creditors offered to clear the debt and pay Hunt $400 if he could come up with a useful invention using a single piece of wire. Three hours later, after much twisting, Hunt had invented the safety pin. His patent specification states that: "The distinguishing features of this invention consist in the construction of a pin made of a single piece of wire or metal combining a spring, a clasp or catch, in which catch, the point of said pin is forced and by its own spring securely retained."

It was something that would still be in everyday use, largely unaltered, more than 150 years later, but, apart from the $400, Hunt did not profit from his invention—to clear the debt he had to assign the rights to the safety pin to his creditors, who are named on the patent as Wm. Richardson and Jno. Richardson (both U.S.).

Left Elizabeth Hurley wears "That Dress" designed by Versace (1994)
Above Ur-punk Johnny Rotten of the Sex Pistols models safety pins (c. 1977)

The inventor: Walter Hunt

1796 Born in Watertown, New York State. Later trains as a mechanic

1827 Submits his first patent, relating to an alarm for buses. Subsequently submits patents for other inventions including a stove, a saw, furniture castors, a knife sharpener, and a precursor of the repeating rifle

1834 Invents the first lock stitch sewing machine, but does not patent it *See also: Sewing machine, page 38*

1849 Patents the safety pin, but has to assign the rights to his creditors, the Richardsons

1859 Files his last patent (relating to a new type of heel for shoes or boots). Dies and is buried in Greenwood Cemetery, New York City

The invention of the zipper in 1913 seemed to provide the ultimate convenient, secure fastening. But George de Mestral thought otherwise and, frustrated by a jammed zipper, he invented the Velcro brand hook & loop fastener in 1941.

ZIPPER/VELCRO

During the 19th century inventors came up with a number of new clothes fastenings including the hook and eye, invented by Charles Atwood (U.S.) in 1843, the snap fastener, invented by Paul-Albert Regnault (France) in 1855, and the press stud, invented by John Newnham (England) in 1860. Snap fasteners were often placed together in rows, and it was a laborious process to fasten and unfasten them all individually, which led Whitcomb L. Judson (U.S.) to invent something that he called a "clasp locker or unlocker for shoes" (patent filed 1891, granted 1893). Judson's invention used a sliding device to lock or unlock a row of clasps and is sometimes acknowledged as the first zipper because, although the zipper itself came 20 years later, it was a direct development of Judson's idea.

With Lewis Walker (U.S.), Judson set up a company to manufacture his fastening, but their developments remained unreliable until, in 1913, company engineer Gideon Sundback (Sweden–U.S.) made the crucial breakthrough. Sundback's ingenious solution was to use metal teeth on flexible tape, together with a slide fastener. First he tried a set of metal jaws on one side closing onto beaded fastenings on the other, and then he conceived the idea of identical sets of interlocking teeth on both sides. He patented the improved idea in 1914 as a "separable fastener"—the zipper was born.

The zipper remained unchallenged until George de Mestral (Switzerland) invented the Velcro brand hook & loop fastener. After walking his dog one day in 1941, de Mestral noticed cockleburs sticking to his clothes and to his dog's coat. Using a microscope, he saw that the cockleburs were covered with tiny hooks that had stuck to loops in the clothes' fabric and dog's fur and, remembering his frustration with a jammed zipper, he wondered if he could use hooks and loops to create a new type of fastener. He filed patents for his idea in 1951 and took the name of his invention from the French *velours croché*, meaning "hooked velvet."

Zipper inventors: Gideon Sundback & Whitcomb L. Judson

1880 Otto Frederick Gideon Sundback is born in Sweden. Little is known about the early life of Whitcomb Judson

1889–92 Judson receives 14 patents relating to street railways and founds the Judson Pneumatic Street Railway Company

1891 Judson files a patent for "clasp locker...for shoes"

1893 Judson and Lewis Walker (U.S.) set up a company to manufacture fastenings

1903 Sundback successfully completes an electrical engineering degree in Germany

1905 Sundback immigrates to the U.S. and works for the Westinghouse Electric Corp.

c. 1908 Sundback is employed by Judson and Lewis's company, now known as the Automatic Hook & Eye Co.

1909 Judson dies, having received 30 patents for fastenings, street railways, internal combustion engines, variable speed transmissions, clutches, and traction wheels

1913 Sundback invents the zipper (patented in 1914), marketed as the Hookless No.2

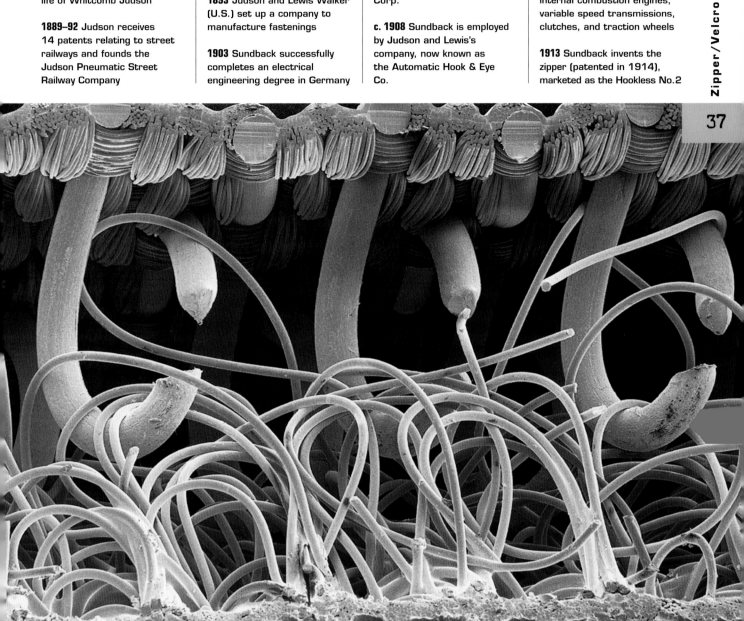

Opposite Zippers employ teeth as interlocking devices **Above** Image of fastened Velcro, taken using a scanning electron microscope. The hooks (top, in pink) interlock with the loops (bottom, in brown) to provide the fastening

Did you know?

In 1851 Elias Howe, Jr. (U.S., inventor of the sewing machine) patented a zipperlike "automatic, continuous clothing closure" described in his patent as: "a series of clasps united by a connecting cord running or sliding upon ribs." He did not manufacture his invention commercially, leaving Whitcomb Judson to reinvent the idea of an automatic closure 40 years later.
See also: Sewing machine, page 38

Gideon Sundback married Elvira, the daughter of Peter Aron Aronson, one of the mechanics at the Automatic Hook & Eye Co. The first British patent for a sliding fastener was taken out in Aronson's name.

Whitcomb Judson's financial backer, Lewis Walker, said of him: "His inventive capacity was great; his practical utility of that capacity was almost nil." Which might explain why it was Gideon Sundback, not Judson, who made the breakthrough that led to a practical zipper.

Zipper inventors: Gideon Sundback & Whitcomb L. Judson

Various people have been named as the inventor of the sewing machine, but early machines could produce only chain stitches. The first patent for a lock stitch sewing machine was granted to Elias Howe, Jr. on September 10, 1846.

The inventor: Elias Howe, Jr.

1819 Born on July 9 in Spencer, Massachusetts, the son of a farmer

1830 Begins working for a neighboring farmer and then works in the mills on his father's farm

1835 Begins working in a machine shop in Lowell, Massachusetts, manufacturing cotton-spinning machinery

1837 Apprenticed to Ari Davis, manufacturer of scientific apparatus, where he hears a customer say that anyone who

SEWING MACHINE

The issue of who should be named as the inventor of the sewing machine is a complex one. The first patent for a sewing machine was issued to Thomas Saint (England) in 1790, in which Saint specified a number of features that would be reinvented more than half a century later by Elias Howe, Jr. and Isaac Singer (both U.S.), but it seems that Saint's machine never progressed beyond a prototype.

A number of patents followed, but the first commercially successful sewing machine was patented by Barthélemy Thimmonier (France) in 1830. Eighty of Thimmonier's chain-stitch machines were used for making French military uniforms, but the machines were soon destroyed by an angry mob of workers fearful for their jobs, and Thimmonier died in penury. Meanwhile, Walter Hunt (U.S.) invented a lock stitch machine c. 1833, the advantage of a lock stitch being that if one stitch breaks the entire chain does not come unraveled. However, Hunt did not patent his invention, and it was left to Elias Howe, Jr. (who was unaware of Hunt's work) to file the first patent for a lock stitch sewing machine more than a decade later.

Howe's wife had taken up sewing to help support their large family, and watching her sew inspired him to produce, in 1846, his first successful sewing machine. He could not find an American manufacturer, but sent his brother Amasa to England, where Amasa sold the manufacturing rights to a corsetmaker named William Thomas. Elias traveled to London to adapt his machine for Thomas, but later fell out with him and returned, penniless, to the U.S. in 1849. He soon discovered that several people had copied his idea, including Isaac Merritt Singer (U.S.), who had been asked to repair one of the copied machines and decided that he could improve it. Howe sued for patent infringement, and in 1854 the courts found in his favor, ruling that Singer should pay royalties to Howe, who subsequently became the fourth richest man in America.

See also: Safety pin, page 34

Above Elias Howe, Jr. (1819–67) patented the first mechanical sewing machine
Opposite Bride and groom admire presents, which include a sewing machine (188

Did you know?

After the courts ruled that Isaac Singer had infringed Elias Howe's patent, Singer patented 19 improvements of his own over the next 13 years and became America's largest producer of domestic sewing machines.

The world's first installment plan scheme was introduced to boost sales of sewing machines. Isaac Singer's business partner, Edward Clark (U.S.), devised a scheme, introduced nationally in September 1856, whereby buyers could take a machine by paying a deposit, and then pay the balance in installments.

Walter Hunt, pioneer of the lock stitch, was also the inventor of the safety pin. The reason that he did not patent or market his lock stitch machine was that he did not want machines to ruin the livelihoods of manual workers.

can invent a practicable sewing machine will be a very rich man

1839 Marries at the age of 20. He fathers several children and his wife takes up sewing to help support the family; watching her sew inspires him to build his first sewing machine

1843 Completes his first sewing machine, which is a failure

1846 Granted the world's first patent for a lock stitch sewing machine and later travels to England to adapt his machine for William Thomas

1849 Returns to the U.S.

1851 Patents an "automatic continuous clothing closure," a precursor of the zipper. Isaac Singer (U.S.) patents the first of his sewing machine improvements *See also: Zipper/Velcro, page 36*

1854 U.S. courts rule that Singer has infringed Howe's patent

1861 Howe is granted a 70-year extension of his patent

1867 Dies on October 3 in Brooklyn, New York, aged 48

The handheld hairdryer was made possible by the diminishing size of electric motors. The first two models went on sale in the U.S. in 1920, and hairdryers developed independently in Germany shortly afterward.

HANDHELD HAIRDRYER

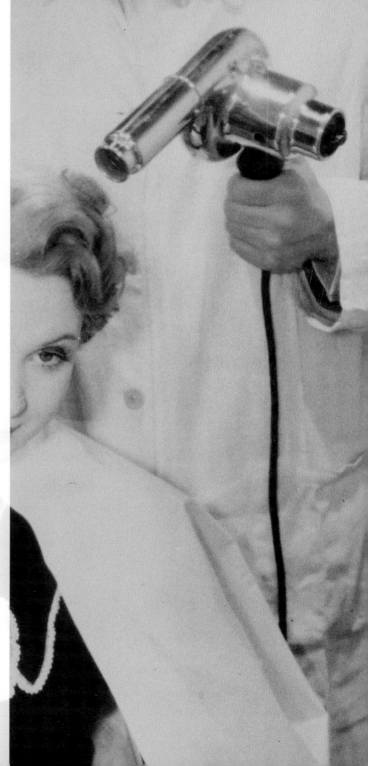

The handheld electric hairdryer is a relatively simple appliance comprising an electric heater and an electric motor driving a fan; the inventive leap came in combining these two existing technologies to make a new product. Electric heaters developed from the invention of the light bulb—it was noted that early filaments converted more energy into heat than light, and this observation led to the development of electric heaters using high-resistance wires as heating elements, mounted on a reflector to radiate heat in one direction.

Making a miniature heater was no problem, so all that was needed to make a handheld hairdryer was an electric motor small enough to fit into a compact casing. Chester A. Beach (U.S.) was a pioneer of small, high-speed, low-power universal electric motors, and made his first such motor in 1905. He later joined forces with Fred Osius and L.H. Hamilton (both U.S.) to form the Hamilton Beach Manufacturing Company, which was based in Racine, Wisconsin. Beach's motors were used in developing several appliances including the first patented electric food mixer in 1910, an electric sewing machine in 1912, and a fixed-stand hairdryer in 1914.

By 1920 electric motors were small enough for Hamilton Beach to produce the Cyclone, one of the world's first two models of handheld electric hairdryer. The other was the Race, produced by Racine Universal Motor Co. in the same town at almost exactly the same time. Hairdryers gradually improved, with smaller, lighter, quieter motors and variable temperature settings and fan speeds, but the next big development, during the 1950s, saw the appearance of the first hairdryers to have plastic casings. Plastic not only allowed for more fashionable designs, it also made hairdryers lighter still and paved the way for the various styling nozzles, volumizers, and diffusers without which today's hairdryers would look like little more than a case containing a heater and a fan.

See also: Light bulb, page 10; Food processor, page 20; Sewing machine, page 38

Did you know?

In 1971 sales of hairdryers dropped dramatically in Britain after TV character Val Barlow was electrocuted by her hairdryer in front of millions of viewers during an episode of the soap opera *Coronation Street*.

The permanent wave, or perm, was invented by hairdresser Karl Ludwig Nessler (Germany, later known as Charles Nestlé). It was a development of the earlier marcel wave, invented by Marcel Grateau (France) in 1872, and was first demonstrated in Nessler's salon on Oxford Street, London, in 1906.

In 1907 Eugene Schueller (France) invented Auréole, a synthetic hair dye whose title later developed into the company name L'Oréal.

Opposite An early electric hairdryer in use in the 1930s **Above** In the 1970s, hairdryers became smaller and more stylish

Hamilton Beach Manufacturing Co. time line

1904 Chester A. Beach (U.S.) joins the U.S. Standard Electrical Works in Racine, Wisconsin. Former cashier L.H. Hamilton (U.S.) joins the company the same year

1905 Beach makes his first high-speed, low-power universal electric motor (known as a fractional horsepower motor)

1910 Together with Fred Osius (U.S.) and L.H. Hamilton, Beach forms the Hamilton Beach Manufacturing Co. The new company produces the first patented electric food mixer

1920 Hamilton Beach produces the Cyclone, one of the world's first two electric handheld hairdryers

1922 The company is sold and eventually becomes part of Glen Dimplex (Ireland)

1933 Fred Osius patents a food blender that later becomes known as the Miracle Mixer and then as the Waring blender

1990 80 percent of Hamilton Beach is sold to NACCO

Industries and merged with Proctor-Silex to form the largest U.S. manufacturer of small kitchen appliances

The Lava lamp was an essential part of 1960s hippiedom and has recently undergone a revival. The original was invented by Craven Walker, who said during the swinging sixties: "If you buy my lamp, you won't need drugs."

The inventor: Craven Walker

1918 Born Edward Craven Walker on July 4 in Singapore

1939–45 Serves as a Royal Air Force (RAF) squadron leader

1953–63 Develops an egg timer into a working Lava lamp. Makes the film *Travelling Light*

1964 David George Smith (England) patents the Lava lamp on behalf of Walker's company, Crestworth Ltd.

1967 Walker himself patents improvements to the Lava lamp

LAVA LAMP

In a country pub during the Second World War, RAF Squadron Leader Craven Walker (England) saw an egg-timer-cum-lamp that he described as "a contraption made out of a cocktail shaker, old tins and things." Primitive it may have been, but it was to be the inspiration for one of the iconic designs of the 1960s—the Lava lamp.

After the war Walker set up a company called Crestworth Ltd. to develop his invention. A light bulb in the base of the lamp would not only provide light, but also heat up a specially formulated wax. The warm wax would rise slowly through a colored liquid, creating dreamy patterns until it cooled sufficiently to sink slowly back again, passing other globules of rising wax as it fell, before repeating the cycle. It sounds simple, but it took ten years to achieve the right balance of oil, wax, and water to create the desired effect. Lava lamps went on sale in 1963, and the patent was filed on March 18, 1964, by David George Smith, on behalf of Crestworth.

The Lava lamp was an immediate success, and was marketed in Europe as the Astro lamp (still used as the name for the "classic" Lava lamp) and in America as the Lava Lite. However, sales dropped during the 1980s and Walker issued a challenge to businesswoman Cressida Granger: If she could make Crestworth profitable within a year she could have the main share of the business. Granger did revive the fortunes of the Lava lamp, and she and Walker formed a new company. Granger later bought out Walker and renamed the company Mathmos, which continues to develop "design-led kinetic lighting products."

Walker's original design is reminiscent not only of the cocktail shaker that inspired it, but also of 1960s space rockets. Today Mathmos announces: "When we invented the Astro lamp in 1963, space travel was in its infancy, but we enjoyed the idea that you could quite possibly commute from Pluto....What happened to the ideals of inter-galactic, planetary commuting? Fight for progression. But remember, Mathmos always propels you into orbit."

Did you know?

Mathmos remains faithful to its hippie roots with helpful comments printed on the packaging, such as: "If you find this box too heavy, don't struggle alone, get help from a kind friend." The message continues: "Part of the Mathmos consumer care regime, helping you (the consumer) consume with utmost confidence in a world of increasing uncertainty."

As well as inventing the Lava lamp, Craven Walker also produced *Travelling Light*, the first naturist film to receive public release in Britain.

1990 Forms a new company with Cressida Granger, which she later renames Mathmos

2000 Dies on August 15 in Ringwood, Hampshire, aged 82, having remained a consultant to and director of Mathmos since 1990

Below Artist's impression of giant Lava lamp building planned for Soap Lake, Washington, February 2003 **Right** The Telstar Lava lamp

AND ALSO...

Lipstick

The sliding-tube container for solid lipstick was invented and manufactured by Maurice Levy (U.S.) in 1915. Five years later, Hollywood film director Preston Sturges (U.S.) invented kissproof lipstick.

Electric kettle

The first electric kettle was manufactured by the Carpenter Electric Manufacturing Co. (U.S.) in 1891, possibly the same kettle as was more famously displayed at the Chicago World's Fair in 1893. This kettle, however, was unreliable, inefficient, and way ahead of its time—most people preferred to boil water over a fire or on a hob. The first electric kettle with a fully immersed element (more than twice as efficient as an external element) was the Swan, invented by Bulpitt & Sons of Birmingham, England, and first marketed in 1921.

Roll-on deodorant

Roll-on deodorants are essentially giant ballpoint pens. Mum (England) invented the roll-on during the 1950s after one of the company's British product developers looked at his ballpoint and realized that the same method could be used to apply liquid deodorant. Test-marketing of the first roll-on deodorant took place in the U.S. in 1952, and it was launched as Ban Roll-On in 1955. Three years later Ban Roll-on was launched in the U.K. as Mum Rollette. (In 1888 Mum had invented a zinc oxide deodorant cream, which was applied with the fingers from a tin.)

Having been inspired by the ballpoint, the roll-on in turn inspired the creation of the computer mouse. Dean Hovey (U.S.), co-designer of the first commercially successful mouse, describes the start of the design process: "Over the weekend I hacked together a simple spatial prototype of what this thing might be, with Teflon and a ball. The first mouse had a Ban Roll-On ball."

See also: Ballpoint pen, page 130;
Computer & mouse, page 128

Spider ladder

On May 11, 1994, Edward Thomas Patrick Doughney was granted a British patent for a: "Spider ladder provided with means for attachment to an item of sanitary ware"—in other words, a ladder to allow hapless spiders to climb out of the bath.

Cotton swabs

Cotton swabs were conceived by Leo Gerstenzang (Poland–U.S.) in 1923, after he had seen his wife carefully wrapping a piece of cotton around the end of a toothpick while caring for their baby. Gerstenzang's invention was perfected in 1925 and originally launched as Baby Gays, a name that was later changed to Q-tips Baby Gays and then simply Q-tips.

Abrasive shaving device

On August 31, 1899, Samuel L. Bligh (U.S.) filed a patent for a shaving device that was doomed to failure. King Camp Gillette (U.S.) had not yet invented the safety razor (he filed his patent in 1901; it was granted in 1904), and the electric shaver, invented by Jacob Schick (U.S.), did not arrive until 1931. An alternative to the cutthroat razor was clearly needed, but Bligh's invention was not the answer. His patent decribed "a device for shaving by abrasion...it consists, substantially, in a roller having an abrading-surface and adapted for connecting with a suitable driving power, whereby the roller may have imparted thereto a rotary motion." In other words, it was a piece of sandpaper on a rotating drum. But the patent office does not take into account the practicality of an invention, and this patent was granted on March 27, 1900.

Baby patting machine

On November 7, 1968, Thomas V. Zelenka (U.S.) filed a patent for an electrically operated mechanical arm that attached to the side of a baby's crib. Zelenka's patent states: "It is generally well known to most parents of small

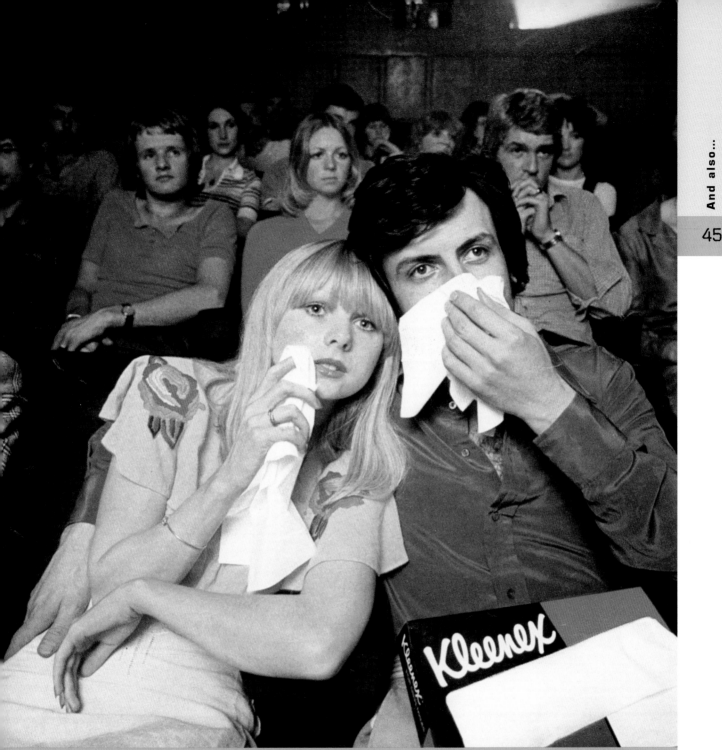

Opposite top *Lipsticks II* (detail), painting by Phillip Le Bas **Above** "Kleenex Tissues for (wo)men in love," undated advertisement (1970s)

infants and children that it is sometimes difficult for the infant to fall asleep, and the parent must resort to patting the baby to sleep by repeated pats upon the hind parts thereof." It goes on to say this invention "will pat a baby to sleep, thereby eliminating the necessity of a parent to do the same manually for an extended period of time."

Paper tissues

Paper tissues are taken so much for granted that it is hard to imagine anyone had to invent them. Founded as a paper mill by John Kimberly and Charles Clark (both U.S.)

in 1872, Kimberly-Clark began to expand and diversify. Just before the First World War the company's chemists invented a new paper product called cellulose wadding, which was used for medical dressings. Looking for other uses for this wadding, Kimberly-Clark launched Kotex sanitary napkins in 1920, and four years later introduced the first paper tissues, which were intended for makeup removal and launched as Celluwipes. Sales were slow until Celluwipes were relaunched as paper handkerchiefs, this time under the name Kleenex-'Kerchiefs—a name later shortened to Kleenex.

Chapter Two

EATING AND DRINKING

After 16 years of determined effort in the face of a series of setbacks that read like the plot of a Gothic novel, Otto Frederick Rohwedder finally succeeded in producing a working version of his invention, the bread-slicing machine.

The inventor: Otto Frederick Rohwedder

18?? Little is known about Rohwedder's early life

1912 Begins work on his first bread-slicing machine

1912–26 Invents and patents a display rack for bread and two unsuccessful bread-slicing

machines that use metal staples to hold the sliced loaf together

1915 Told that he has one year to live

1917 Loses all his tools and equipment in a fire

SLICED BREAD

A jeweller by trade, Otto Frederick Rohwedder (U.S.) had something of an obsession with bread, filing more than a dozen patents relating to the production and sale of sliced and unsliced loaves. He began working on his first bread slicer in 1912, and when he asked bakers whether they thought it was a good idea, they told him no, because the bread would go stale too quickly, so he expanded the scope of his invention to slice *and wrap* the bread.

His first major setback had nothing to do with the inventive process—in 1915 his doctor told him that he had only a year to live. Two years later, already defying this gloomy prediction, a second disaster befell him when a fire in his workshop destroyed his prototype and all his tools. A lesser man might have taken this as a sign that bread was not intended to be sold presliced, but Rohwedder persevered, secured financial backing in 1922, and produced a working bread slicer in 1928.

For many years it was generally accepted that the first presliced loaf of bread went on sale in Battle Creek, Michigan, that same year, but in 2001 journalist Catherine Stortz Ripley of the *Chillicothe Constitution Tribune* in Missouri discovered an article in the paper's archives that suggested otherwise. Under the headline Sliced Bread Is Sold Here, the paper reported on July 6, 1928: "Chillicothe Baking Co. the First Bakers in the World to Sell this Product to the Public," and presliced loaves reportedly went on sale in Chillicothe the following day. In 2003 Rohwedder's son Richard told Stortz Ripley: "My father and Frank Bench [the owner of the bakery] were friends. When no-one else in the world would give my father's machine a try, Frank Bench did....Other bakers scoffed at the idea." They may have scoffed, but they soon had to eat their words—by 1933 an estimated 80 percent of bread sold in America was presliced.

1922 Secures financial backing for his bread-slicing machine

1928 Perfects a machine for slicing and wrapping bread. The first presliced loaves go on sale, in Chillicothe, Missouri

1928–35 Files various patents

for handling and processing loaves of bread, including four for improvements to his bread-slicing machine

1960 Dies on November 8

Above Machine-sliced bread has made Dagwood-style sandwiches possible **Below** Two employees of the Excelsior Baking Company operate a bread-slicing machine, Minneapolis, Minnesota (c.1928)

Did you know?

The 1928 news article in the *Chillicothe Constitution Tribune* announcing the first sale of sliced bread explained: "The idea of sliced bread may be startling to some people. Certainly it represents a definite departure from the usual method of supplying the consumer with bakers loaves. As one considers this new service one cannot help but be won over to the realization that here indeed is a type of service which is sound, sensible and in every way a progressive refinement in bakers bread service."

In August 2003, at a special ceremony to mark the 75th anniversary of the invention of sliced bread, Otto Rohwedder's son Richard was presented by the mayor of Chillicothe with a golden key to the city in a presentation case marked with a plaque reading: "Chillicothe, the home of sliced bread."

The concept of breakfast cereal began with Shredded Wheat, invented as an aid to digestion by Henry D. Perky in 1893. His idea was soon expanded upon when John Harvey Kellogg invented flaked breakfast cereal in 1895.

50 BREAKFAST CEREAL

Because he suffered badly from dyspepsia, lawyer Henry D. Perky (U.S.) did not always live up to his name, often being far from perky. One morning at breakfast in a hotel in Nebraska, he met a fellow dyspeptic who recommended eating whole boiled wheat with milk for breakfast, something that not only improved Perky's digestion, but inspired him to invent the first ready-to-eat breakfast cereal, Shredded Wheat, in 1893. His invention proved so popular when sold locally that in 1895 Perky founded the Natural Food Co. and began manufacturing Shredded Wheat commercially.

That same year, John Harvey Kellogg (U.S., known as J.H.) filed a patent for: "Flaked cereals and process of preparing same." Kellogg, a qualified doctor, ran the Battle Creek Sanitarium in Michigan, where he promoted what he called "biologic living," a regime that included treatments such as calisthenics, cold baths, enemas, and electroshock therapy. Kellogg was also interested in diet, and often experimented with boiling cereal grains as an aid to digestion. One night in 1894, J.H.'s brother Will Keith (U.S., known as W.K.) left some wheat grain overnight between boiling the grain and rolling it out into dough. The grains became moist, and, when he rolled them the next morning, they formed flakes instead of the usual flat sheet of dough.

Fortunately for the future morning routine of millions, instead of throwing away the flakes, W.K. baked them and served them for breakfast. The patients enjoyed them, but although it was W.K. who accidentally invented flaked cereal, it was J.H. who filed a patent the following year, beginning: "Be it known that I, John Harvey Kellogg…have invented a certain new and useful Alimentary Product and Process of Making the Same." Tension between the two brothers eventually came to a head when W.K. added malt flavoring to a cereal without J.H.'s approval, resulting in W.K. buying out his brother and founding the Battle Creek Toasted Corn Flake Company, eventually known simply as Kellogg's.

Flaked cereal inventors: John Harvey Kellogg & Will Keith Kellogg

1852 J.H. Kellogg born on February 26 in Tyrone, Michigan, the son of a farmer

1860 W.K. Kellogg born on April 7 in Battle Creek, Michigan

1875 J.H. graduates from Bellevue Hospital Medical College

1876 J.H. takes over the running of the Battle Creek Sanitarium

1894 W.K. accidentally discovers a process of creating cereal flakes

1895 J.H. announces the invention of Granose Flakes in the February issue of *Food Health* magazine, and files a patent for: "Flaked cereals and process of preparing same" on May 31 (granted April 14, 1896)

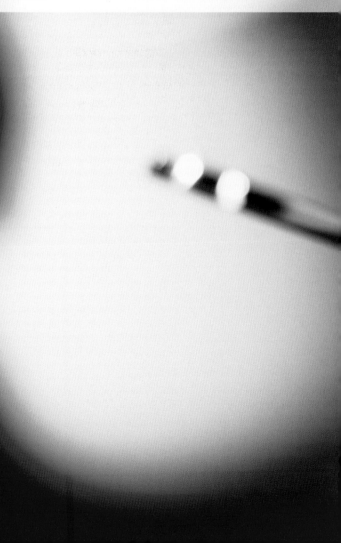

Did you know?

Puffed rice was first introduced at the St. Louis World's Fair in 1904, where it was sold as a snack because it was seen as a competitor to popcorn—but its potential as a breakfast cereal was soon realized.

Right John Harvey Kellogg on his 86th birthday (February 26, 1938) **Below** A brimming bowlful of cornflakes

1898 The brothers apply their process to corn and introduce Corn Flakes

1906 W.K. buys out J.H. and founds the Battle Creek Toasted Corn Flake Company (later W.K. Kellogg Co., later Kellogg's)

1943 J.H. dies on October 6 in Battle Creek, aged 91, having filed patents for various foodstuffs and for nonfood inventions including radiant-heat baths, inhalers, and body braces

1951 W.K. dies on October 6 in Battle Creek, aged 91, having established the W.K. Kellogg Foundation, one of America's leading philanthropic institutions

Roman soldiers were issued licorice root to chew, and it has been used in confectionery since the 17th century, but Licorice Allsorts were not invented until 1899, when licorice salesman Charlie Thompson dropped his tray of samples.

Bassett's (confectioners) time line

1818 George Bassett born in Ashover, near Chesterfield, England

1832–39 Apprenticed to a confectioner and fruiterer

1842 Establishes his own business in Sheffield, England,

as a "Wholesale Confectioner, Lozenge Maker and British Wine Dealer"

1851 Employs S.M. Johnson (England) as an apprentice

1893 Johnson becomes sole proprietor of Bassett's (the

LICORICE ALLSORTS

Licorice itself is not an invention, but a plant (*Glycyrrhiza glabra*) whose roots contain glycyrrhizin, a substance that in its pure form is 50 times sweeter than sugar. Licorice, whose name is a corruption of the original Greek *glycorrhiza,* meaning "sweet root," has been noted for its sweet taste and healing properties since ancient times, and cultivation began on a large scale in Europe during the 16th century, not least in Pontefract, England. Used as a flavoring and coloring for foodstuffs ranging from gingerbread to beer and stout, licorice is most commonly known for its use in confectionery, although 90 percent of the world crop is now used for flavoring tobacco.

The oldest known licorice sweet is the Pontefract Cake, also known as the Pomfret Cake or Yorkshire Penny, which is known to have been made as early as 1614 and is still made in the Yorkshire town to this day. In 1760 Pontefract confectioner George Dunhill (England) took the first step toward the invention of Licorice Allsorts when he had the idea of mixing sugar and flour with licorice to make the familiar, colorful "liquorice and paste" sweets. Then, in 1899, salesman Charlie Thompson (England), of sweet manufacturer Bassett's, accidentally invented Licorice Allsorts when he dropped his tray of individual samples in front of a prospective customer and they scattered all over the counter; seeing the colorful assortment they made when mixed together, the customer immediately placed an order for a mixed bag, and "allsorts" were born.

Traditionally, Licorice Allsorts comprised Chips, Rocks, Buttons, Cubes, Nuggets, Plugs, and Twists. Then, in 1999, Bassett's introduced two new Licorice Allsorts to mark the centenary of Thompson's invention. One was a checkered oblong Licorice Allsort made from two strips of licorice with two strips of strawberry paste in a checkered pattern, and the other was a replica of Bassett's mascot, Bertie Bassett, made from aniseed and blueberry.

Did you know?

In Yorkshire, licorice is known colloquially as spanish, presumably a reference to the fact that when demand outstripped supply in the 19th century licorice began to be imported from Spain. A more fanciful story is that a Pontefract schoolmaster found some sticks washed up on the beach after the defeat of the Spanish Armada and took them back to Pontefract to use for caning his pupils. The pupils would put splinters from the sticks in their mouths to stop them crying out in pain, and soon discovered that the "Spanish" sticks tasted sweet, leading to the birth of licorice confectionery—and, no doubt, to an upsurge in caning offenses at the school.

Former poet laureate John Betjeman (England) celebrated love and licorice in his poem "The Licorice Fields at Pontefract":

In the licorice fields at Pontefract
My love and I did meet
And many a burdened licorice bush
Was blooming round our feet;
Red hair she had and golden skin,
Her sulky lips were shaped for sin,
Her sturdy legs were flannel-slack'd
The strongest legs in Pontefract

Bassett's also invented Jelly Babies, which were launched as Peace Babies to celebrate the end of the First World War. Rationing meant that production ceased during the Second World War, but when they were reintroduced in 1953 the name was changed to Jelly Babies.

Johnson family remains part of Bassett's until 1974)

1899 Charlie Thompson (England), a Bassett's salesman (collectively known as "Bassett's Travellers"), drops his tray of samples while visiting a customer in Leicester, England, accidentally inventing Licorice Allsorts

1926 Bassett's becomes a public company and commissions Greenlys advertising agency to design a trademark

1929 On January 1 Bassett's registers Bertie Bassett, the Licorice Allsort man, as a trademark

1961 Bassett's takes over licorice-makers Wilkinson's of Pontefract, Yorkshire

1990 Bassett's becomes part of Trebor Bassett, a division of Cadbury Schweppes PLC (Britain)

1999 The company introduces two new Licorice Allsorts to mark the centenary of Thompson's invention

Below A jumble of Licorice Allsorts

TEA BAG/INSTANT COFFEE

Tea bag

The first patent for a tea bag was issued to A.V. Smith (England) in 1896, but the fact that nothing is known about him is a reflection of how unsuccessful his invention proved to be, probably because the English preferred more traditional methods of brewing tea. Some eight years later, c.1904, tea and coffee merchant Thomas Sullivan (U.S.) began supplying tea samples in silk envelopes rather than tins and his customers discovered that the tea could be brewed in these "bags," but it was another 15 years before the first purpose-made tea bags were introduced by Joseph Krieger (U.S.). Initially, tea bags were used only by caterers, but by the 1930s the majority of tea bags in the U.S. were bought for use in the home, where they took the place of the traditional perforated metal "tea egg" or "tea ball," which held the tea leaves while they infused.

Instant coffee

In 1930, hoping to stimulate flagging sales of coffee beans and reduce excess stocks, the Brazilian Institute of Coffee asked food companies to develop ways of producing an instant coffee that was soluble in water and yet retained its flavor (previous attempts had been totally unpalatable). Scientist Max Mortgenthaler of Nestlé (Switzerland) began researching ways of drying percolated coffee and invented a technique known as spray drying. Brewed coffee was sprayed into the top of a heated tower, the droplets of coffee dried out as they fell and by the time they reached the bottom they had become a powder. Mortgenthaler perfected his technique in 1937 and the following year Nestlé launched Nescafé as the first successful instant coffee.

Morgenthaler's original method was later improved by further processing the powder into granules, and in 1964 Nestlé invented the superior, but more expensive, technique of freeze-drying, launching Nescafé Gold Blend in 1965 as the first freeze-dried instant coffee.

Tea bag & instant coffee timeline

1896 A.V. Smith (England) invents and patents the first tea bag

1903 Dr Satori Kato (Japan–U.S.) invents and patents a process for manufacturing soluble coffee powder, but it is not marketed commercially

c.1904 Tea and coffee merchant Thomas Sullivan (U.S.) sends samples of tea in silk envelopes, and customers discover that the tea can be brewed in the bags

1918 Benjamin Hirschorn (U.S.) files the first U.S. patent for a tea bag

c.1919 Joseph Krieger (U.S.) introduces the first purpose-made tea bags

1930 The Brazilian Institute of Coffee asks food companies to find new coffee products as a way of reducing excess stocks

1930s Tea bags, previously used only by commercial caterers, are first used in American homes

1937 Nestlé scientist Max Mortgenthaler perfects a technique for spray-drying coffee

1938 Nestlé launches Nescafé, the first instant coffee, and takes out a patent for instant tea

1953 The Tetley Tea Group introduce tea bags to Britain

1965 Nestlé launches Nescafé Gold Blend as the first freeze-dried instant coffee

1991 Nestlé launches Nescafé Cappuccino as the first instant cappuccino

Did you know?

In 17th-century Constantinople (now Istanbul, Turkey) it was traditional at weddings for the groom to provide the bride with coffee, and vow to always do so—failure to continue providing coffee was considered grounds for divorce.

Ex-U.S. First Lady Nancy Reagan once said: "A woman is like a tea bag. It's only when she's in hot water that you realize how strong she is."

Through the ages some religious fanatics have sought to ban coffee, claiming it was the invention of Satan, but during the 17th century Pope Clement blessed coffee in order to make it a drink "fit for Christians."

Opposite top Coffee production in Guatemala: mother and grandmother look on as this Guatemalan girl prepares a cup of instant coffee. Most of Guatemala's coffee is earmarked for export **Below** Thrift means reusing your tea bags

At one time glass was too delicate to put in an oven, but the invention of heatproof glass by German experts in the late 19th century, and its subsequent development in America as Pyrex cookware, brought about a culinary revolution.

Corning Glass timeline

1851 Amory Houghton, Sr., (USA) switches his business from trading in coal, wood, and other materials to glass. He buys Cate & Phillip (later Bay State Glass), Union Glass Works, and, later, the Brooklyn Flint Glass Works in Brooklyn, New York

c.1868 Houghton and his son move their company to Corning, New York, and change the company name to Corning Glass

1912–15 Heatproof borosilicate glass is developed by J.T. Littleton, Eugene C. Sullivan,

OVENPROOF GLASS

Today we take ovenproof glassware for granted, but at one time the idea of placing a glass dish in an oven would have seemed incredibly foolish. Even after the invention of a suitable heatproof glass, the idea of using it for cooking seemed so improbable that Pyrex ovenware came about almost by accident.

Toward the end of the 19th century, Otto Friederich Schott, Carl Zeiss, and Ernst Abbe (all Germany) established a specialist glassworks called the Glastechnische Versuchsanstalt at Jena, Germany. Their factory became renowned for the production of high-quality lenses, microscopes, binoculars, and other optical equipment. It later became the Jena Glassworks of Schott & Associates and, from 1919, part of the Carl Zeiss Foundation. In 1884 the trio developed a new type of heat- and chemical-resistant glass, known as borosilicate glass from its main ingredients, boron oxide and silica.

Then, in 1912, J.T. Littleton, Eugene Sullivan, and William Taylor (all U.S.) of Corning Glass developed the Germans' borosilicate glass into an improved heatproof glass known as Nonex, which at first was used purely for industrial purposes. Legend has it that the idea of using this glass for ovenware came about in 1913, when the wife of one of the Corning technicians baked a cake in a battery jar made from Nonex. As a result, the company was inspired to develop the first ovenproof glass for domestic use, which was launched in 1915 under the trademark Pyrex.

Not only was the glass inventive, but so were the designs—a 1937 advertisement for a Pyrex casserole dish announced: "It's like getting three dishes in one!" The advertisement went on to explain how the lid and the base could be used as separate dishes or used together as: "The Complete Casserole. Modern, streamlined, practical...You can see the food cooking in 'Pyrex' Brand dishes."

and William C. Taylor of Corning Glass (all U.S.), based on the 1894 invention of Otto Friederich Schott, and marketed under the trademark Pyrex

1926 Corning Glass invents a fully automatic machine to blow electric light bulbs

1931–38 The Owens-Illinois Glass Company (now Owens-Illinois, Inc.) and Corning Glass develop fiberglass, based on the 1836 invention of M. Dubus-Bonnel (France)

1947 Photosensitive glass is developed by S.D. Stookey (U.S.) of Corning Glass

1964 Photochromic glass is developed by W.H. Armistead (U.S.) and S.D. Stookey of Corning Glass

1967 High-strength laminated sheet glass, suitable for dinnerware, is developed by James W. Giffen (U.S.) of Corning Glass and marketed under the trademark Corelle

1970 The first communications-grade fiber optics are developed by Donald Keck, Robert Maurer and Peter Schultz (all U.S.) of Corning Glass

Ovenproof glass

57

Below A lampworker at the Kontes Glass Company in Vineland, New Jersey, heats a laboratory jar (July 1992). The Kontes plant uses only Pyrex borosilicate glass in the manufacture of laboratory glassware

Did you know?

Without Corning Glass, Thomas Edison (U.S.) might not have been able to develop the first commercially practicable light bulb—in 1879 the glass for Edison's famous lamp was blown by company glassblower Fred Douerlein (U.S.). *See also: Light bulb, page 10*

British rugby league club Castleford Tigers was formerly known as "The Glassblowers" in honor of what was once a major industry in the town.

The earliest of William Painter's 85 patented inventions related to rail travel, but his most lasting invention was the bottle cap—the crimped metal bottle stopper that is still in use more than 110 years after it was invented.

BOTTLE CAP

Did you know?

In his patent, William Painter suggested a list of household tools for opening his seal—"a knife, a screwdriver, a nail, an icepick"—but he later went on to invent a bottle cap opener as well as patenting several improvements to the bottle cap itself.

Although the bottle cap was a hugely successful idea, its inventor, William Painter (U.S.), is possibly less famous for this than he is for inspiring King Camp Gillette (U.S.) to invent the safety razor. In 1855, at the age of 17, Painter began a four-year apprenticeship at a leather manufacturer, during which time he filed patents for his first two inventions, a fare box that was capable of giving change and a railway carriage seat that converted into a couch. He went on to patent a counterfeit-coin detector in 1862 and a kerosene lamp burner in 1863, before taking a job in the machine shop of Murrill & Keizer (U.S.) in 1865, where he was to work for the next 20 years, during which time he invented and patented more than 35 industrial tools and devices.

But Painter's greatest invention was the bottle cap. Bottling fizzy drinks was a problem in the 19th century, because the volatile contents would often expel a conventional cork. In 1880 Painter set his mind to solving this problem and in April 1885 he was granted a patent for a reusable swing stopper known as the Triumph. In September that year he was granted a patent for a single-use stopper, which, being one-tenth the cost of the Triumph, was an immediate success but made his earlier invention redundant. Still not satisfied that he had invented the best solution, Painter went on to invent the bottle cap in 1891, an invention so successful that it is still in use, essentially unchanged, more than 110 years later.

Modern salesmen often talk about the "razor-blade principle," or the "Gillette principle," whereby a disposable product generates its own market for replacements. But, although the principle refers to Gillette and his invention, it was not Gillette who thought of it—as an aspiring inventor, Gillette was working as a salesman for Painter, and it was Painter who advised him: "Invent something that will be used once and then thrown away. Then the customer will come back for more."

Left A beer bottle float, complete with cap, at the Reading Bicentennial in Reading, Pennsylvania (1948) **Above** A bevy of caps

The inventor: William Painter

1838 Born on November 20 in Triadelphia, Maryland, the son of a farmer. He is later educated at schools run by the Society of Friends (Quakers) in Fallston, Maryland, and Wilmington, Delaware

1855 Begins a four-year apprenticeship at patent leather manufacturers Pyle, Wilson & Pyle (U.S.) in Wilmington

1858 Granted patents on August 3 for a fare box and on August 31 for a railway carriage seat

1862 Granted a patent on July 8 for a counterfeit-coin detector

1863 Granted a patent on June 30 for a kerosene lamp burner

1865–85 Works for Murrill & Keizer (U.S.) and patents

35 inventions including a seed sower, a soldering tool, telephonic equipment, and a number of pump valves

1885 Files a patent for the Triumph swing stopper (granted April 14) and forms the Triumph Bottle Stopper Company. June, files a patent for a single-use stopper (granted September 29) and reorganizes his company as the Bottle Seal Company

1891 Invents and files a patent on May 19 for the bottle cap (granted February 2, 1892)

1892 Reorganizes his company again on March 9, this time the Crown Cork & Seal Company

1903 Retires with 85 patents to his name

1906 Dies on July 15 in Baltimore, aged 67

Did you know?

One predecessor of the bottle cap was a marble stopper patented by Hiram Codd (England) in 1872—the pressure of the gas in the bottle forced the marble up against a rubber gasket to form a seal. Sometimes the stopper did not work, giving rise to the term "codswallop" for bad beer ("wallop" was slang for beer), a term that later came to mean "nonsense."

At the end of the 20th century the ring pull appeared in *Time* magazine's list of "one hundred great things," but the ring pull per se, with a ring and a pull tab, was merely the second in a long line of tear-strip openers for drinks cans.

RING PULL

Canned food was introduced in 1812, but cans were not used for drinks until the 1930s. Early drink cans were made of steel and were opened with a "church key," which pierced a triangular hole in the top of the can; a second hole was required on the opposite side as a vent in order for the drink to flow. In the late 1950s steel cans were superseded by aluminum cans, which were easier to open but still needed a separate opener. Then, in 1959, Ermal Fraze (U.S.) found himself at a picnic with no church key; he used a car bumper to open his beer before reportedly uttering the inventor's phrase: "There must be a better way."

Some time later, reputedly in an effort to lull himself to sleep on a caffeine-induced sleepless night, Fraze began to investigate the idea of inventing an integral lever for a drink can, to negate the need for a separate opener. The idea was simple—score a tear strip into the can top and provide a lever to open and remove it. The engineering was not so easy, but, as the founder of the Dayton Reliable Tool & Manufacturing Company, Fraze had the necessary metalworking expertise, and in 1963 he was granted a patent for an: "Ornamental design of a closure with a tear-strip opener."

Fraze's ornamental design had an approximately rectangular tab (useful as a lever, but difficult to grip for the pull) and a keyhole-shaped tear strip. The big breakthrough came in 1965, when Omar Brown and Don Peters (both U.S.) improved on Fraze's design by introducing a ring-shaped tab (into which a finger could be inserted) and by simplifying the shape of the tear strip, ideas that they patented on behalf of Fraze. The ring pull performed its function perfectly, but it also created large amounts of sharp litter. Then, in 1973, Brown took the idea a step further and invented the precursor of the "push-in fold-back" inseparable tear strip (again patented on behalf of Fraze, granted 1975), thus solving the litter problem while retaining the convenience and ease of opening.

See also: Can opener, page 74

Left A participant in the Earth Day celebration in New York, wearing a tabard made from ring pulls that he recycled for the occasion (April 1980) **Above** Smile! Tear-strips as art: photograph by Markku Lahdesmaki

The inventor: Ermal Fraze

1913 Born Ermal Cleon Fraze ("Ernie") in Indiana. Educated at Ohio Northern College and later at the General Motors Institute in Flint, Michigan, from which he graduates as a mechanical engineer

1949 Establishes his business, Dayton Reliable Tool & Manufacturing Company

1959 Uses a car bumper to open a can of beer at a picnic and conceives the idea of the tear-strip top

1963 Granted a patent for the tear-strip top

1965 Omar Brown & Don Peters (both U.S.) file a patent on July 6, on behalf of Fraze, for a "ring-shaped tab for tear strips of containers" (granted October 31, 1967)

1975 Brown is granted a patent on behalf of Fraze for a "can end with inseparable tear strip"—the push-in fold-back tear strip, first introduced in 1977

1987 Fraze receives the Distinguished Alumnus Award from the General Motors Institute, which has been awarded only five times previously: to three presidents and two vice presidents of General Motors

1989 Dies in Dayton, Ohio, aged 76

Did you know?

In 1990 Robert DeMars and Spencer Mackay (both U.S.) were granted a patent for an invention that overcame the fact that, according to DeMars' and Mackay's patent, "push-in fold-back" openers can be difficult to open "by older people or people with arthritis or other afflictions." The inventive leap was a so-called camming protruberance that sticks up from the top of the can. The tab is rotated onto the protruberance, lifting the end high enough to get a finger under it and, at the same time, breaking the seal to start the "push-in" part of the operation.

The reason that drink cans have a stepped neck, making them narrower at the top, is because the thicker steel required for the top is the most expensive part of the manufacturing process: The stepped neck reduces the size of the top by about 20 percent.

Once upon a time draft beer in cans was a contradiction in terms. Now, thanks to the invention of the widget by two scientists working for the Guinness Brewery, that formerly self-contradictory phrase is a dream come true for beer drinkers.

WIDGET

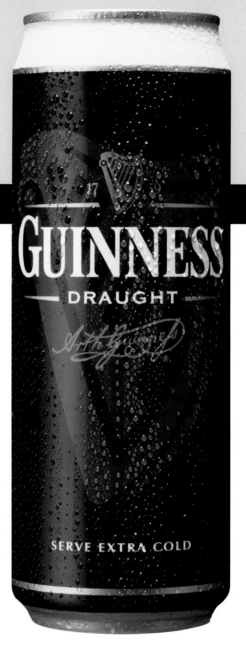

The Guinness Brewery (Ireland) is justifiably proud of the thick creamy head that forms the characteristic white band at the top of its famous stout. But until 1988 the head appeared only on draft Guinness, as pulled in pubs and bars, not on bottled or canned Guinness. To overcome this drawback, Guinness began to investigate ways of re-creating the taste and texture of draft stout in bottles and cans.

For draft Guinness, "sparklers" on the pumps create the creamy head by adding small amounts of nitrogen to the stout, and this concept formed the basis of several attempts to re-create a head from a bottle or can. Two early Guinness patents described three methods of adding carbon dioxide and nitrogen to the drink: by pouring it over polystyrene granules, by using an ultrasonic generator, or by injecting it with a syringe. All three methods worked, but, as acknowledged in the widget patent, the need for extraneous devices to create the head was a big disadvantage, so the hunt was on for a device that could be integrated into the can or bottle.

Then, on November 29, 1985, after Guinness had spent four years and £5 million on research, Alan James Forage (England) and William John Byrne (Ireland) filed a patent on the company's behalf for a small plastic disk known as the in-can system (ICS). More popularly known as the widget, the ICS is a small nitrogen-filled capsule that has a tiny hole in it. Because the beer is pressurized with carbon dioxide, the nitrogen is forced to stay in the widget, but when the can is opened the pressure drops and the widget releases the nitrogen into the beer. This forces the beer to release dissolved carbon dioxide, which, with the nitrogen, creates a head on the beer. The ICS first became available in 1988 and won a Queen's Award for Technological Achievement in 1991. Since then the widget has been continually improved, and in 1999 Guinness invented the "rocket" widget for use in bottles.

Widget time line

1985 Alan Forage (England) & William Byrne (Ireland) invent the widget (ICS), for which they file a patent on November 29, on behalf of the Guinness Brewery

1988 Draft Guinness in a can is test-marketed as the first beer to use a widget

1989 Draught Guinness in a can is launched nationally in Britain

1991 The widget wins a Queen's Award for Technological Achievement

1996 Guinness initiates Project Snake—the development of a floating widget

1997 Guinness introduces the floating widget to draft Guinness in cans

1999 The "rocket" widget is launched, enabling consumers to drink draft Guinness straight from the bottle

2003 Readers of the technical innovation website T3.co.uk vote the widget the greatest technological invention of the past 40 years

Did you know?

The Guinness Brewery was founded in 1759, when Arthur Guinness (Ireland) traveled to Dublin with £100, which he had been left in his godfather's will. He bought a disused brewery at St. James's Gate, on a 9,000-year lease, and turned it into what is now one of the biggest breweries in the world.

Other brewers have since invented their own version of the widget, and the draft treatment was applied to canned and bottled bitter in 1992, lager in 1994, and cider in 1997.

The trademark of the O'Neill harp and Arthur Guinness's signature was introduced in 1862. The harp gave its name to Harp lager, first sold in Ireland in 1960 and Britain in 1961.

Left A can of Guinness stout **Below** Did this creamy head come from a can?

Known as the wine cask in Australia, where it was invented by Thomas Angove, the wine box succeeded in changing Australia's drinking habits and has also made significant inroads against the bottle in the rest of the wine-drinking world.

WINE BOX

The wine box, also known variously as the soft pack, bag-in-a-box or wine cask, is a very simple invention that is said to have been inspired when Thomas Angove (Australia) saw a picture of a Greek shepherd drinking from a goatskin. Wine spoils when exposed to air, so Angove decided to keep the air from the wine by putting the wine in a bag—the bag collapsed as the wine flowed out, protecting the remaining wine from the air. He then went a step further than the ancient Greeks had done by placing the soft bag in a rigid cardboard container, making it more practical to stack, store, and transport. The orginal one-gallon pack announced: "This package combines the ancient method with modern technology. Inside the carton, the wine is contained in a specially formulated inert plastic 'skin'."

Like so many inventions, the idea was simple, but the execution was not. Angove came up with the idea of the wine box in 1963, but it took two years to develop a tap and a suitable membrane for the bag, and it was not until April 20, 1965, that he was granted a patent for his "improved container and pack for liquids." In September of that year Angove's company, Angove's Winemakers & Distillery Pty. Ltd., launched his invention, selling one-gallon wine boxes of Angove's white wine, red wine, port, sweet sherry, and muscat in both Australia and England.

However, there were problems. Unlike glass, the membrane of the bag was permeable to oxygen, which meant that "bags of wine," as they were at first known, had a limited shelf life. In 1971, having proved that the wine box had great market potential, Angove's stopped producing wine in boxes because of permeability and other technical problems. Other winemakers following Angove's lead had similar problems, but technology gradually improved, and in 1984 Angove's reentered the wine box market with the Paddle Wheel 5—a 5 quart (5 liter) box of a wine whose name commemorated the company's early days when wine was taken down the River Murray to Murray Bridge by paddle steamer.

Below A water-seller in Tangier, Morocco, rings a bell to attract customers. He carries a goatskin water bag and several brass drinking cups
Above Thomas Angove

The inventor: Thomas Angove

1917 Born Thomas William Carlyon Angove on August 8 in Renmark, South Australia, and later educated at Holdfast Bay Prep School, Wykeham Prep School, and St. Peter's College

1936 Enrolls for a diploma in Agriculture at Roseworthy Agricultural College

1938 Transfers to the Oenology (study of wine) course at Roseworthy

1940 Graduates with an R.D.Oe., gaining all prizes awarded for the course: the Leo Buring Gold medal for highest aggregate work, the R.H. Martin Tasting Prize, and the Karl Weidenhofer Prize for Individual Study for his project on brandy distillation. (This project is later published in booklet form and subsequently reprinted)

1941 Becomes a director of the family business, Angove's Winemakers & Distillers Pty. Ltd. Enlists in the Royal Australian Air Force as a trainee pilot and serves until discharged in September 1944 with the rank of flight lieutenant

1945 Elected a member of the Federal Viticultural Council of Australia (ultimately the Australian Wine and Brandy Producers' Association Inc.)

1946 Becomes managing director of Angove's

1963 Conceives the idea of the wine box

1965 Granted a patent on April 20 for an "improved container and pack for liquids." Wine is sold in wine boxes for the first time that September

1967 Elected a member of the Council of the Australian Wine Research Institute

1969 Begins developing what is the largest single vineyard in Australia at the time, with 473 hectares planted with 23 different grape varieties. Pioneers vineyard and trellising techniques that will be suitable for mechanical harvesting, which is in its infancy and will come to revolutionize Australia's grape-growing industry

1977 Awarded the Queen Elizabeth II Silver Jubilee Medal for services to the wine industry

1983 Becomes chairman of Angove's and is succeeded as managing director by his son John

1993 Invested as a Patron of the Wine Industry of Australia in honor of his outstanding contribution to the affairs of the Australian Wine Industry

1994 Appointed a Member in the General Division of the Order of Australia (AM) in the Australia Day Honors List

2001 Retires as chairman of the board of directors of Angove's

Until the mid-20th century milk was sold in bottles or decanted into open containers. That all changed with the invention of Tetra Classic, a tetrahedron-shaped carton that was to revolutionize the way liquid foodstuffs were packaged.

TETRA PAK

In 1929 Ruben Rausing (Sweden) co-founded a packaging company specializing in packaging dry foodstuffs such as flour, sugar, and salt. Then, in 1943, he broadened the scope of the business into liquid foods and initiated the development of his concept for a cardboard milk carton. Selling milk in cartons may seem obvious today, but in Rausing's time it was unheard of and required new techniques for coating paper with plastic to make it fluidproof, and for hygienically sealing the packs. However, realizing the concept proved not as easy as conceiving it because, as Rausing observed later: "Doing something that nobody else has done before is actually quite hard."

Company researcher Erik Wallenberg (Sweden) conceived the idea of tetrahedron-shaped cartons, Rausing invented a process of forming and filling them from a continuous tube of treated paper, and Rausing's wife suggested sealing them by placing sealing clamps on the tube while it was full, so that no air reached the contents of each pack. The result was the carton now known as Tetra Classic, for which a patent was filed on March 27, 1944.

Rausing and Wallenberg founded AB Tetra Pak in 1951, and on May 18 that year, after more than six years developing a suitable means of coating the paper, they announced to the press the invention of their new tetrahedral carton. The following September the first Tetra Pak machine was delivered to the local dairy, and from November that year cream was packaged in 3 ounce, (100 milliliter) Tetra Pak cartons. Rausing and Wallenberg's revolutionary new pack was an immediate success, being hygienic, cheap to manufacture, and convenient to stack, store, and transport. Tetra Pak may have started life as the first milk carton, but, more than 50 years after its launch, the company's products have spread far beyond the dairy and are now used to package all manner of liquid foods and drink, including soup, fruit juices, tomato paste, ice-cream bars, fruit compôtes, and even wine.

Left Milk cartons speed along conveyor belts (2000) **Above** Musicologist Harry Smith pouring milk, photographed by poet Allen Ginsberg (January 1985)

The inventor: Ruben Rausing

1895 Born on June 17 in Raus, near Helsingborg, Sweden

1918 Graduates with first class honors from the Stockholm School of Economics and Business Administration, Sweden

1920 Graduates from Columbia University in New York with a master's degree in economics

1929 Establishes the packaging company Åkerlund & Rausing in Lund, Sweden, with Erik Åkerlund (Sweden)

1933 Buys out Åkerlund so that he owns the company outright

1943 Initiates development of a revolutionary new milk carton

1944 Embraces the idea of researcher Erik Wallenberg (Sweden) for a tetrahedron-shaped carton, and applies for a patent on March 27

1946 Åkerlund & Rausing produces a prototype filling machine, designed by Harry Järund

1951 Rausing founds AB Tetra Pak with Wallenberg in Lund, Sweden, initially as a subsidiary of Åkerlund & Rausing

1961 AB Tetra Pak produces the first aseptically filled and sealed carton (for UHT milk)

1983 Rausing dies on August, 10, aged 88

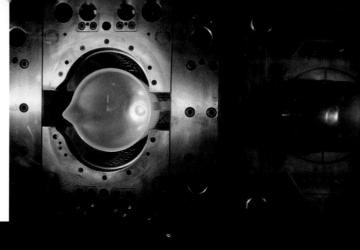

The famous Tupperware range of resealable plastic food containers was named after its inventor, Earl Silas Tupper, but the range also owes much of its success to Brownie Wise, the woman who conceived the idea of the Tupperware party.

TUPPERWARE

Life without plastic food containers seems inconceivable now, but when Earl Silas Tupper (U.S.) was a chemist working for DuPont (U.S.) during the 1930s, the storage choice was in glass jars, pottery containers, or tins. Tupper had the foresight to think that plastics were the materials of the future (though not, at first, for food containers), and in 1938 he left DuPont to set up the Tupper Plastics Company and begin his own research into new plastics.

Tupper's particular interest was the production of plastics from waste materials, and, having maintained a good relationship with DuPont, he asked the company for materials with which to experiment. DuPont gave him a quantity of a hard, black plastic known as polyethylene slag, an oil-refining by-product, from which Tupper succeeded in producing a tough, flexible, moldable plastic. He then had to think of a product that could be made from this new plastic. He is said to have considered shoe heels, but it was food containers, first produced in 1945, with which he literally made his name. The inventive leap, and the aspect of Tupperware he was able to patent (filed 1947, granted 1949), was the airtight seal, said to have been modeled on an upturned paint tin lid.

Despite the advantages of being airtight, spillproof, and virtually unbreakable, Tupper's containers did not sell well at first, partly because plastic was seen as an inferior material, but mainly because consumers did not have the knack of "burping" the container to expel excess air and make the airtight seal. The "burping" problem precipitated a social revolution in 1951 when sales representative Brownie Wise (U.S.) suggested demonstrating the technique to groups of potential customers in their homes. This wise idea soon became known as the Tupperware party and was so successful that Wise was appointed vice president of Tupperware Home Parties, and Tupper took the bold step of removing his products from the shops and selling them purely through home demonstrations.

See also: Polythene, page 252

Did you know?

Tupperware has been exhibited at a number of art and design museums including the Museum of Modern Art in New York, and the Smithsonian Institution in Washington, D.C.

Brownie Wise, whose sales techniques earned her a position as vice president of Tupperware Home Parties, the sole distributor of Tupperware, often used the catchphrase "If we build the people, they'll build the business."

It is estimated that at the turn of the millennium, a Tupperware party started somewhere in the world every three seconds.

Opposite top A single-impression mold used in the injection molding of a Tupperware Mix-n-stor container, shown with a container sitting in the moulding casing (1965) **Below** Tupperware advertisement showing a Tupperware party in progress (1960s)

The inventor: Earl Silas Tupper

1907 Born on July 28 in Berlin, New Hampshire, the son of a farmer who invented various labor-saving devices

1928 Takes a course in tree surgery and sets up a tree surgery and landscaping business, Tupper Tree Doctors

1937 Begins working for DuPont (U.S.), where, he says later, "my education really began"

1938 Leaves DuPont to form Tupper Plastics Co., subcontracting a lot of work from DuPont

1945 Produces his first plastic consumer products, including cigarette cases and an unbreakable drink tumbler

1947 Files a patent on June 2 for: "Open mouth container and nonsnap type of closure therefor" (granted November 8, 1949)

1951 Registers the trademark Tupperware in the U.S. Brownie Wise (U.S.) conceives of parties as a development of a sales technique she used as a representative for Stanley Home Products

1958 Retires after selling his company for several million dollars

1965 The trademark Tupperware is registered in Britain

1983 Tupper dies on October 5, aged 76

A merchant seaman who abandoned the sea to pursue a yen for invention, Luther Childs Crowell patented many inventions relating to paper. His method of making square-bottomed grocery bags, patented in 1872, is still in use today.

The patentee: Luther Childs Crowell

1840 Born in West Dennis, Cape Cod, Massachusetts, the son of a sea captain

1862 Granted patent for his first invention, an "aerial machine"

1867 Granted a patent on May 28 for a method of incorporating metal into the neck of a paper bag for ease of filling and sealing. Invents a stove polishing brush

1872 Granted patents on February 20 for the square-bottomed grocery bag and machinery for producing bags

GROCERY BAG

Luther Childs Crowell (U.S.) grew up with a fascination for paper, and it is said that his neighbors would often see him spending hours at a time folding pieces of paper into various shapes and designs. His first invention, however, had nothing to do with paper (he filed a patent in 1862 for an "aerial machine"), but he soon turned his attention to the field that had fascinated him as a child and in 1867 he was granted a patent for the use of thin strips of metal in the neck of a bag, to hold it open while it was being filled, and then to provide an easily re-sealable closure.

Five years later he was granted a patent for the square-bottomed paper bag that is still widely used across the U.S. for groceries (although plastic bags are more common in Europe). Crowell was not the first to think of a square-bottomed grocery bag—that accolade goes to Margaret Knight (U.S., aka "the mother of grocery bags"), who in 1869 invented what she called "flat or satchel-bottomed bags"—but in 1872 he was granted a patent for inventing a method of manufacturing such bags that is still in use more than a century later. Crowell's patent stated: "I am aware that paper-bags have been made which will assume a quadrangular shape when filled; but claim the method herein described as the most simple and practical." The patent office agreed, and granted him two patents, one for his method and one for the machinery to mass-produce such bags from a long tube of folded paper.

A decade later Crowell invented a machine that would fold newspapers as they came off the press, something first used by the *Boston Herald*. Two years after that he began working for R. Hoe (U.S.), a printing company that had inadvertently infringed his patents, but, realizing the potential of his ideas, had then paid him royalties on the infringed patents and employed him to come up with more inventions for the print and newspaper trade.

1877 Granted a patent for his invention of a machine to fold newspapers

1879 Begins working for the printing company R. Hoe (U.S.), subsequently inventing paper-feeding machinery and machinery for wrapping and folding newspapers

1892 Granted a patent for a seal for newspapers to prevent fraudulent returns of unsold papers by newsagents. (Unscrupulous newsagents would often collect used copies of papers and return any that were in good enough condition to pass as unsold: Crowell's seal prevented this because if

the seal was broken it proved that the paper had been read)

1906 Dies in his birth town of West Dennis, aged 66, having filed nearly 300 patents

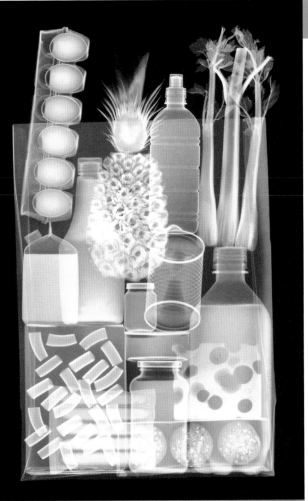

Left and Above The square-bottomed grocery bag, inside and out

Did you know?

According to *Guinness World Records*, the world's largest collection of paper and plastic bags belongs to Heinz Schmidt-Bachem (Germany). Since 1975 he has collected 150,000 bags ranging from those supplied by corner shops to those bearing the names of designer boutiques. The pride of his collection is the world's first industrially manufactured paper bag, which dates from 1853.

Most famous as the inventor of the Anywayup Cup, Mandy Haberman is also the inventor of the Anyware range of baby-feeding products. Her first invention, the Haberman Feeder, was inspired by the needs of her daughter, Emily.

ANYWAYUP CUP

In 1980 Emily Haberman, daughter of graphic designer Mandy Haberman (England), was born with Stickler's syndrome, a congenital condition in which a child suffers from, among other things, an abnormally small lower jaw, a posterior cleft palate and an abnormally set-back tongue. When Emily was two, Mandy began designing a special feeding device for babies and children who suffered from such problems, and in 1984 prototypes of the Haberman Feeder were tested both on healthy babies and those with feeding problems. First sold by mail order in 1987, the Haberman Feeder continues to be sold throughout the world to specialist units, hospitals, and clinics.

In 1990 Haberman conceived the phenomenally successful invention that was to become a household name: the Anywayup Cup. After watching a friend's child spilling drinks over the carpet and furniture, she set her mind to designing a genuine no-spill cup, and invented a unique slit valve, which she patented in 1992. In its final form, the valve forms an integral part of the mouthpiece, ensuring that drink flows only when the baby or child sucks, and that the Anywayup Cup lives up to the marketing boast that children can shake, rattle, and roll it without spilling a drop.

Sales of the Anywayup Cup rocketed in 1997 after Haberman persuaded supermarket chain Tesco to stock it. She recalls: "We realized that we needed to get the product into the supermarkets, but this was not going to be easy....Then we had a brainwave. We had nothing to lose, so we took a chance. We filled a cup with concentrated Ribena [deep purple juice] and put it loose inside a white box and posted it to the buyer! A few days later, the telephone rang—we were into Tesco!" By the turn of the millennium, sales had passed seven million per year and Haberman was fêted as a celebrity inventor, being named British Female Inventor of the Year 2000 and nominated for a Special Recognition Award from the Global Women Inventors and Innovators Network in 2003.

Did you know?

Haberman writes of her difficult experiences in trying to sell the idea of the Anywayup Cup: "At one company I chucked a cup full of juice onto the Director's desk, right on top of all his papers. Thank God, it didn't spill a drop! He was impressed, once he'd recovered from the shock. However, his enthusiasm eventually waned when he realized that he wasn't going to get something for nothing....Now, instead of having to interest companies in the product, I have to sue them for infringement. But I guess that is the price you have to pay for success. We are now a name to be reckoned with in the industry. The man whose desk I chucked the cup onto has been in touch with us several times since trying to persuade me to grant him a licence. Sorry, mate—no chance!"

Opposite below "More milk please—and next time I want it in an Anywayup Cup" **Above** Mandy Haberman demonstrating the Anywayup Cup

The inventor: Mandy Haberman

1954 Born Mandy Brecker on October 19 in Hertfordshire, England, and later educated at Enfield County School for Girls and then Hornsey and St. Martins School of Art, in London, where she gains a degree in graphic design

1980 Her daughter Emily is born with Stickler's syndrome

1982 Begins to research and develop a feeder for babies with sucking problems

1984 Produces the first prototypes of the Haberman Feeder

1987 Sets up her own company and begins production of the Haberman Feeder

1990 Conceives the idea of a leakproof trainer cup

1992 Files the first of many patents to protect her idea of a slit valve to control the flow of liquid

1994 Produces final prototypes of the Anywayup Cup

1995 Together with V&A Marketing Ltd. (Wales), Haberman launches the prototype at two baby and toddler fairs, securing £10,000 worth of advance orders

1996 The Haberman Company Ltd. is set up to supply the Anywayup Cup. In addition, Haberman grants a manufacturing and distribution license to The First Years Inc. (U.S.), which markets the Anywayup Cup in the U.S. under the brand name Tumble Mates

1997 Sebastian Conran Associates (England, now Conran & Partners) is appointed to redesign the Anywayup Cup range

1999 Sebastian Conran Associates designs the Anywayup Smiley Cup, developed in conjunction with the U.K. Health Education Authority to coincide with its dental health campaign "Smile"

2000 Haberman is named British Female Inventor of the Year at a ceremony organized by the Professional Family Women's Network. The Anywayup Cup wins a Gold Medal at the Salon International des Inventions in Geneva, Switzerland, a Horners' Award for innovative use of plastics (sponsored by the British Plastics Federation), Best Consumer Product and the 3M Award for Innovation at the Design Effectiveness Awards 2000, and the Tommy Award for Most Parent Friendly Innovative Product (sponsored by St. Thomas's Hospital, London)

2001 Appointed a member of the Chartered Institute of Patent Agents (CIPA) Disciplinary Board, the Intellectual Property Strategic Information Technology Advisory Committee (IPSITAC), and the Intellectual Property Advisory Committee. Invited to judge the recently renamed International Design Effectiveness Awards (IDEAs) and the *Tomorrow's World* Awards (sponsored by the BBC innovations program *Tomorrow's World* and NESTA, the National Endowment for

Science, Technology and the Arts)

2002 Becomes a member of the Advisory Council for the European Commission's Information Society Technologies Programme (PRISM). Launches an online intellectual property rights discussion forum called Make Sparks Fly (www.makesparksfly.com). Awarded an Honorary Doctorate in Design and made an honorary Fellow of Bournemouth University, England

2003 Nominated for a Special Recognition Award from the Global Women Inventors and Innovators Network (GWIIN), whose stated goal is to "promote, recognize and celebrate the achievements of women inventors and innovators while raising awareness of the positive contribution that women bring to the economy"

The first patent for canned food was issued to Peter Durand in 1810, but the can's useful little companion, the can opener, was not invented until nearly half a century later —until then, people had to use weapons or a hammer and chisel to open cans.

CAN OPENER

Necessity is often said to be the mother of invention, but in the case of the can opener it was a long gestation—canned food was patented in 1810 and first produced in 1812, but an opener was not invented until 1855, by Robert Yeates (England). At first, cans were used mainly for military rations, and personnel would use swords, bayonets, or even gunfire to open them. A can of meat supplied to the Royal Navy in 1824 carried the instruction: "Cut round on top with a chisel and hammer." The situation did not improve even when canned food became available to the public c. 1830—people were advised to open cans using household tools, and shopkeepers would often open them for customers in the shop, rather defeating the object of the food being canned.

Ezra Warner is often credited with inventing the can opener but in 1855 cutler and surgical instrument maker Robert Yeates filed a patent for: "Improvements applicable to lock knives and lever knives, partly applicable to such surgical and other instruments as may be connected to handles by moving joints." Essentially, he had invented an improved version of what today would be called a penknife, with "blades" that folded into the handle—these blades included a knife, fork, spoon, and "a lever knife for cutting or ripping open preserved provision cases etc., such lever knife being a curved blade with a shoulder ...forming an efficient bearing or fulcrum in use." Today, "preserved provision cases" are better known as cans, and Yeates's "lever knife," which he stated was "of novel form and construction," is better known as a can opener.

The first can opener to use a cutting wheel rather than a straight blade was invented in 1870 by William Lyman (U.S.), but it was difficult to use—the exact center of the can had to be pierced with one blade and the width of the opener adjusted to the size of the can before working the wheel round the rim. Lyman's invention was a huge step forward, but the modern rotary can opener, which grips, cuts, and pulls itself around the rim, did not appear until the 1920s.

Above In classic wedding tradition, empty tin cans on a string clatter behind hopeful newlyweds **Below** A rotary can opener in action

Can openers time line

1804 Nicolas Appert (France) invents and tests a method of sealing food in glass jars and heating them to preserve it; the method later becomes known as appertization, after the inventor

1810 Peter Durand (England) is granted a patent for the use of "vessels of tin or other metals" for "preserving animal and vegetable food." He later sells the patent rights to Bryan Donkin & John Hall (England)

1812 Donkin & Hall establish the world's first cannery

1820s Cans carry the instruction "Cut round on top with a chisel and hammer"

1830s People are advised to use household tools to open cans

1855 Robert Yeates (England) invents and patents the first can opener

1858 Ezra Warner (U.S.) invents a two-bladed can opener that is later described as "part bayonet, part sickle," although Warner himself claims: "A child may use it without difficulty, or risk"

1866 J. Osterhoudt (U.S.) invents a can with a key fixed to the top for opening

1870 William Lyman (U.S.) invents the first can opener to use a cutting wheel. (Lyman goes on to patent various inventions, including butter dishes, tea and coffee pots, and a lever for opening jars)

c. 1925 Lyman's invention develops into the modern form of rotary can opener

1927 The wall-mounted can opener is invented by the Central States Manufacturing Company of St. Louis

1970 William Cookson of Cookson Sheet Metal (England) is granted a patent for a tear-open can

Did you know?

Canning food was made possible by the prior invention of a heat-treatment technique for food by Nicolas Appert (France). "Appertization," as it was known, clearly worked, but it was more than half a century before anyone was able to explain why. Then, during the 1860s, Louis Pasteur (France) showed that food spoils due to the action of microorganisms—Appert's heat treatment killed any existing microorganisms, and sealing the food in jars (and, later, cans) prevented contamination by any new ones.

Nonstick pans are often said to be one of the by-products of the space race, but in fact Teflon was invented in 1938 and the means of utilizing it for nonstick pans was invented in 1954, both years before mankind ventured into space.

NONSTICK PAN

Teflon, the coating that makes pans nonstick, was invented serendipitously by Roy Plunkett (U.S.) in 1938. Plunkett was attempting to create a nontoxic refrigerant and thought that he had prepared a container of tetrafluoroethylene gas—but when his assistant, Jack Rebok (U.S.), took the lid off the canister there was no gas inside. The weight of the canister proved that there was *something* in there, so they sawed it in half and discovered a greasy white powder, which proved to have amazing properties: it was not affected by heat or electricity, was not corroded even by the strongest acids and solvents, and had the lowest coefficient of friction of any substance yet discovered or manufactured.

Plunkett had created a polymer known as polytetrafluoroethylene, or PTFE, and on July 1, 1939, he filed a patent for PTFE and his method of producing it (granted 1949). The patent stated that PTFE would be useful for "handling certain corrosive agents, such as hydrofluoric acid, or for protecting workers from the fumes that arise from such reagents." Its first practical use was in clothing as the only material capable of protecting workers on the Manhattan Project (1942–45) from the corrosive uranium hexafluoride used to make uranium-235 for the first atomic bomb. PTFE was kept a military secret until 1946, when it was made public and the trade name Teflon was registered.

The patent also stated that PTFE could be "molded and spun and put to a wide variety of uses where its peculiar properties would be advantageous," but it was not until 1954 that anyone thought of using its low coefficient of friction to make nonstick pans. Engineer Marc Grégoire (France) and his wife, Colette, invented a process for fixing a thin layer of Teflon to metals such as aluminum, and Colette suggested applying it to kitchenware. Grégoire founded the Tefal company in 1955 and produced the first nonstick pans in 1956. Since then Teflon has been used in clothing, engineering, electronics, medicine—and the space program.

The inventors: Roy J. Plunkett & Marc Grégoire

1906 Marc Grégoire born in France

1910 Roy J. Plunkett born on June 26 in New Carlisle, Ohio

1932 Plunkett graduates in chemistry from Manchester College in Ohio

1933 Plunkett receives a master's degree in 1933 from Ohio State University

1936 Plunkett receives a Ph.D. in organic chemistry from Ohio State University. Joins E.I. DuPont de Nemours & Co. (U.S.) as a research chemist at the Jackson Laboratory in Deepwater, New Jersey

1938 On April 6 Plunkett and his assistant, Jack Rebok (U.S.), open a cylinder of tetrafluoroethylene to discover that Plunkett has by chance created polytetrafluoroethylene (PTFE, or Teflon)

1939 Plunkett files a patent on July 1, on behalf of Kinetic Chemicals, a subsidiary of DuPont (both U.S.), for "tetrafluoroethylene polymers." Becomes a chemical supervisor at DuPont's Chambers Works at Deepwater

1946 Teflon is registered as a trademark in the U.S.

1954 Teflon is registered as a trademark in Britain. The Grégoires invent and patent a method of bonding Teflon to aluminum (U.K. patent filed by Colette Grégoire & Georgette Wamant in 1955, granted 1959)

1956 Marc Grégoire founds Tefal in Sarcelles, France. Tefal launches the first nonstick pans

1975 Plunkett retires from DuPont

1988 DuPont honors Plunkett with an award in his name, first awarded to celebrate the 50th anniversary of the invention of Teflon. The Plunkett Award recognizes those who contribute important new products using Teflon

1994 Plunkett dies on May 12, aged 83

1996 Grégoire dies aged 90

2000 Tefal invents and launches Thermo-Spot, the first heat indicator to be built into the nonstick coating of a frying pan

Did you know?
Former President Ronald Reagan was known as the Teflon President because no accusations or criticisms regarding him or his policies seemed to stick.

Did you know?

Teflon has far more uses than as a coating for nonstick pans. As well as its initial use to protect clothing and equipment against corrosives, it is also used in electronics for insulating cables and components; in medicine for artificial joints and blood vessels (because it is inert and therefore the body does not reject it); in clothing as a stain repellent; and in engineering as a weather-resistant and flame-retardant coating for bridges, buildings, and monuments including the Statue of Liberty, New York, and the Millennium Dome, London.

Opposite top Before Teflon, nearly everything stuck to the pan **Below** "Choose a pan like you choose your man. It's what's inside that counts," UK advertisement for DuPont Teflon coated pans (early 1970s)

Over thousands of years, forks evolved from knives, and spoons developed independently. Then, during the 20th century, a parallel set of utensils was invented to complement the traditional knife, fork, and spoon: the spork, foon, and knirk.

CUTLERY

The spoon is often cited as the first eating utensil, partly because the knife originated as a tool and/or weapon long before it was used for eating food. Shells are assumed to have been the first spoons, their shortcomings when used for scooping liquid from a bowl leading to the later addition of a handle. It is thought that spoons were subsequently carved from wood to obviate the need for a separate handle, a theory given credence by the fact that the word spoon derives from the Anglo-Saxon *spon,* meaning a splinter or chip of wood.

Forked agricultural tools have been in use for centuries, and there is evidence that the ancient Greeks, Romans, and Turks used smaller ceremonial forks, but the development of the table fork evolved from the use of the knife for eating. By the Middle Ages diners had progressed from the use of one knife (used to cut the food and then stab the offcut to transfer it to the mouth) to the use of two: one for holding the food and the other for cutting and stabbing. Because the food would often twist around the knife being used to hold the food, a forked tool developed, with two tines to hold the food more securely, and from this evolved the modern table fork, with three and then four tines.

When four-tined forks were introduced to America they were often called "split spoons," and 20th-century America saw the invention of a variation on this theme: the spork, which combined the bowl of a spoon with the tines of a fork and is now most often seen in fast food outlets. There are conflicting theories as to the origins of the spork, but an official patent on the name and concept was granted to the Van Brode Milling Co., Inc. (U.S.) in 1970. The spork in turn gave rise to the foon, an inverted version created by pressing on the bowl of a plastic spork until it turns inside out. One inventor, known as Crackriot (U.S.), even proposed a knirk—a fork with a sharp blade between the tines, and a safety device to prevent diners from cutting off their lips.

Cutlery time line

Antiquity Spoons develop from the use of shells. Knives with pointed tips come to be used for eating food. The ancient Romans, Greeks, and Turks use ornamental forks

c. 7th century Table forks are thought to have been used in the Middle East

c. 11th century The Italians introduce table forks to Europe

1307 At the death of Edward I, King of England, his inventory refers to seven forks, including one of gold

1380 At the death of Charles V, King of France, his inventory refers to 12 silver and gold forks, some of them decorated with precious stones, with a note that they were used for eating "mulberries and other foods likely to stain the fingers"

1630 Governor Winthrop of the Massachusetts Bay Colony introduces the first fork to America. (The norm at the

Did you know?

Some experts contend that the spork is a direct descendant of the runcible spoon, a name coined by Edward Lear in *The Owl and the Pussycat* and since used to describe a type of pickle fork. It also appears in Tony Peek's song *Slipping Away*:

I'm slipping away my own true love,
I'll be gone by the time you wake up.
I'm slipping away for a year and a
day, with a runcible spoon and a cup.

After visiting America in 1842, Charles Dickens wrote that he saw people "thrust the broad-bladed knives and the two-pronged forks further down their throats than I ever saw the same weapons go before, except in the hands of a skilled juggler."

Opposite top French artist Gaston Palmer poises eight spoons in the air during an act on the Côte d'Azur, France (May 1961)
Below Cutlery and a plate adorn the side of a drinking fountain at Monserrate Palace in Sintra, Portugal

time, according to historian George Francis Dow, was that "knives, spoons and fingers, with plenty of napery, met the demands of table manners")

18th–19th century As two-tined table forks become more common, the pointed tips of knives (no longer used for spearing food) are replaced with bulbous tips to assist the scooping up of food. Straight tines begin to be replaced by curved ones. Three- and four-tined forks are introduced; this, together with curved tines, leads to the use of the fork for scooping food and the development of knives with straight blades rather than bulbous tips

1907 Elmer Walter (U.S.) invents and files a patent for a mirror incorporated in the handle of a knife, fork, or spoon (granted 1908). His patent states: "The user of the implement may have ready at hand a mirror for the purpose of inspecting the teeth in the mouth or the mouth or other portions of the face generally, at any time desired by the user of the implement"

1937 Constance Winifred Honey (England) invents and patents a spoon "made of toffee, chocolate or sufficiently solid candy or sweet stuff to hold a portion of medicine, oil, medicinal or other food" in order to encourage children to take their medicine

1970 Van Brode Milling Co., Inc. (U.S.) is granted a patent for the name and concept of the spork, although it is thought either to have been invented in the 1940s by the American military in Japan, or to have evolved in the

1960s with the introduction of fast food

2002 Crackriot (U.S.) invents the concept of the knirk, a fork with a sprung blade at the midsection, acting like a guillotine at the press of a button, but active only when the knirk is vertical, to prevent diners cutting off their own lips

AND ALSO...

Liquor flask

In September 1885 Herbert William Torr Jenner (Britain living in U.S.) filed a patent for "improvements in liquor flasks" (granted November 1885). The improvement was that the liquor flask was concealed within a case made to look like a book, presumably to allow covert drinking. Jenner's patent states that there would be "a hole in the said covering beneath the bottom of the flask so that it can be pushed upwards, and the end made to project through a hole in the covering." He goes on to claim that although there had been flasks with ornamental coverings before, "in no case has the ornamental covering been so made as to entirely cover and conceal the flask from observation, and at the same time admit of ready access to its contents." Given that Jenner was a patent lawyer, and the title of the "book" covering the flask in his patent drawing was *Legal Decisions, Vol. II*, one wonders whether the intention of his invention was to alleviate tedium in court.

Beer-barrel hat

Patented under the more general title of "Substance Dispensing Headgear," the beer-barrel hat was invented by Randall D. Flann (U.S.) and granted a U.S. patent on October 19, 1999. Flann's patent describes his invention thus: "A headgear for dispensing a substance [that] has a container to carry the substance. A spigot is secured to the container. The spigot can be opened to dispense the substance by gravity, suction, pressure or levity flow.... The spigot can be closed to retain the substance in the chamber. A hat-like recess is formed within the bottom wall of the container sized for wearing on an individual's head, and for maintaining the container in a freestanding condition during hands-free ambulation of the individual."

Pyrex

Pyrex cookware was developed by Corning Glass (U.S.) from 1912 to 1915, and was based on a heat-resistant borosilicate glass invented by glassmaker Otto Friederich Schott in 1884. Schott, Carl Zeiss, and Ernst Abbe (all Germany) had established a specialist glassworks called the Glastechnische Versuchsanstalt at Jena, Germany, which became renowned for the production of high-quality lenses, microscopes, binoculars, and other optical equipment. It later became the Jena Glassworks of Schott & Associates and, from 1919, part of the Carl Zeiss Foundation. In 1912 J.T. Littleton, E.C. Sullivan, and W.C. Taylor (all U.S.) of Corning Glass developed Schott's borosilicate glass into an improved heatproof glass known as Nonex, which at first was used purely for industrial purposes. In 1913 the wife of a Corning technician is said to have baked a cake in a battery jar made from Nonex. This led to the introduction of Pyrex in 1915 as the first ovenproof glass for domestic use. Not only was the glass inventive, but so were the designs—a 1937 advertisement for a Pyrex casserole dish announced: "It's like getting three dishes in one! The Complete Casserole. Modern, streamlined, practical.... The Lid. Turn it over and it becomes the very dish for a fruit tart, baked eggs, dozens of things....The Bottom Part. Used without the lid it is ideal for things like shepherd's pie, macaroni cheese and bread-and-butter puddings. You can see the food cooking in 'Pyrex' Brand dishes."

Mayonnaise

There are various theories as to the origins of mayonnaise. One is that it originated in Bayonne, France; another that it was specially invented for the Duke of Mayenne (France) in 1589, and a third, now generally accepted as the correct one, is that it was introduced to France from the Balearic Islands by the Duc de Richelieu (France), who seized Port Mahon, the capital of Minorca, on June 28, 1756. The story goes that Richelieu, a renowned gastronome, organized a feast to celebrate the taking of the port, and at the feast he was introduced to a local salad dressing called *mahonnaise*.

Opposite top A woman demonstrates the use of a Prohibition-era book that conceals a liquor flask (February 1927) **Above** U.S. marines stationed in London, England, during the Second World War make the most of their mashed potato (August 1942)

Instant mashed potato

The first so-called "instant" mashed potato was produced in the form of flaked or powdered potato, to which the consumer added butter and hot water or milk; however, this was not a very palatable product and was not a huge success. Then, between 1961 and 1962, Edward A. Asselbergs (Canada) of the Department of Agriculture in Ottawa invented dehydrated potato flakes. This form tasted so much better than the earlier products that Asselbergs is often hailed as the inventor of the concept of instant mashed potatoes. In fact, what he had invented was a vast improvement of an existing concept and a food-processing technique that has since been used to provide nutrient-rich foods for deprived countries.

Chapter Three

GETTING AROUND

Apart from being a painter, sculptor, architect, and engineer, Leonardo da Vinci was also a prolific inventor. Some of his inventions were so far ahead of their time that they were not realized until the 20th century.

The inventor: Leonardo da Vinci

1452 Born on April 15 in Vinci, Italy, the illegitimate son of a Florentine notary and a young peasant woman called Caterina

1482 Enters the service of Ludovico Sforza (Italy)

1483–97 Designs an armored vehicle, a parachute, an ornithopter, a glider, a helicopter, and a human-powered aeroplane
See also: Glider, page 106

1497 The first "miter" lock gates, reputedly invented by

LEONARDO'S INVENTIONS

As his name suggests, Leonardo was born in the town of Vinci, between Pisa and Florence in Italy. He was sent to study painting under Andrea del Verrocchio (Italy) c. 1470, and in 1482 he moved to Milan under the patronage of Duke Ludovico Sforza (Italy) who, with the monks of Santa Maria delle Grazie, commissioned *The Last Supper.* After the duke's fall from power, da Vinci moved to Florence, where he found a new patron, Cesare Borgia, Duke of Romagna (Italy), and in 1503 he returned to Florence, where he painted the *Mona Lisa.* As da Vinci had worked for King Louis XII of France, Louis's successor, François I, granted him a yearly allowance from 1516 and allowed him the use of Château Cloux, near Amboise, where da Vinci lived until his death on May 2, 1519.

During the Renaissance there was little separation between the disciplines of art and science, and it was Leonardo's artistic skills that led to his inventive, scientific, and engineering prowess. For him, observation was the key to knowledge, and he applied his painterly eye to discovering the workings of everything about him. While under the patronage of Sforza, he not only painted some of his greatest artworks, but also invented a number of military machines and drew designs for a parachute, an ornithopter, a helicopter, a glider, and a human-powered airplane.

More than 500 years later, Steve Roberts (England) built Leonardo's fixed-wing glider in accordance with the original drawings, and it was successfully flown in October 2002, by British hang-gliding champion Judy Leden. Two years earlier Katarina Ollikanen (Sweden) had built da Vinci's rigid pyramidal parachute, and on June 26, 2000, skydiver Adrian Nicholas (England) proved that it worked, making a safe descent from 10,000 feet (3,000 meters) over South Africa, and saying afterward: "It took one of the greatest minds who ever lived to design it, but it took 500 years to find a man with a brain small enough to actually go and fly it."

da Vinci (although this is disputed), are completed to his designs on the Naviglio Interno, a canal close to Milan. Earlier lock gates had closed vertically and were difficult to seal, but da Vinci's design, with the V of the miter pointing upstream, used the pressure of the water to seal the gates

1502 Becomes senior military architect and general engineer to Cesare Borgia

1506 Employed by Louis XII of France

1508 Studies anatomy in Milan

1509 Surveys and draws maps of Lombardy and Lake Isea

1513 Designs a form of astronomical magnifier, as recorded in his notebook: "In order to observe the nature of the planet, open the roof and bring the image of a single planet onto the base of a concave mirror. The image of the base will show the surface of the planet much magnified"

1516 Awarded an annual allowance by François I of France

1519 Dies on May 2 near Amboise, France, aged 67

Below left Adrian Nicholas descending over Mpumalanga, South Africa, in a modern realization of Leonardo's 15th-century design for a parachute (2000) **Below right** Self-portrait drawing of Leonardo da Vinci (c.1519), one of history's most prolific inventors

LEONARDO VINCI

Did you know?

Leonardo's knowledge of human nature was clearly as great as his knowledge of science. He proposed the idea of a submarine, but did not disclose the design "on account of the evil nature of men who would practise assassinations at the bottom of the seas by breaking the ships in their lowest parts and sinking them together with the crews who are in them." *See also: Submarine, page 104*

Leonardo's last words are reported as: "I have offended God and mankind because my work did not reach the quality it should have."

Ask most people whom they think built the first railway locomotive and they will probably say George Stephenson. But, in fact, the railway locomotive was invented by Richard Trevithick and first demonstrated in February 1804.

RAILWAY LOCOMOTIVE

Richard Trevithick (England) was born in the heart of the Cornish tin-mining area. His father was a mine manager, and Trevithick duly became a mining engineer, devoting his talents to the improvement of the steam engine. At the time James Watt (Scotland, known as the father of steam) had a virtual monopoly on stationary steam engines, but Trevithick demonstrated that steam could be used at far greater pressures than Watt had envisaged, a considerable advance in steam technology that led directly to the development of steam locomotives.

Toward the end of the 18th century Trevithick invented the Cornish Engine, a double-acting, high-pressure engine that worked by cutting off steam early in each stroke to make use of the expansion of the high-pressure steam. In 1801 he produced the first successful steam carriage for use on the road, but the poor state of the roads meant that steam carriages did not fulfill their potential and he did not develop the idea further. Railways, with their iron tracks, offered a better opportunity for steam power, and in 1804 Trevithick built the world's first railway locomotive by mounting one of his stationary steam engines on the chassis of a railway wagon. He made several demonstrations in February 1804, during which his locomotive successfully hauled 10 tons of freight and a number of passengers along the 9½-mile Pen-y-Darren ironworks railway in Wales.

Trevithick's locomotive *Catch Me Who Can* was set up as a demonstration on a circular track near Euston, London, in 1808; a broken rail ended this venture, and Trevithick turned his attention elsewhere, leaving it to others, including George Stephenson (England), to develop and commercialize the railways. Despite Trevithick's inventive mind, he failed to capitalize on his ideas, eventually losing all his wealth in a silver-mining venture in Peru. He was bought a ticket home by George Stephenson's son, Robert, and died penniless in England in 1833.

The inventor: Richard Trevithick

1771 Born on April 13 in Illogan, Cornwall, the son of a mine manager

1796 Builds his first scale model of a steam road carriage

1800 Builds the first of his Cornish Engines

1801 Produces the world's first successful steam road carriage, capable of about 8 mph (13kph)

1803 Produces a second, more powerful steam road carriage, which is demonstrated on a 10-mile (16 km) route in London

1804 Demonstrates the world's first railway locomotive

1807 Takes over supervision of the Thames Tunnel, but the river breaks in the following year and work is abandoned (eventually completed in 1842)

1808 Exhibits *Catch Me Who Can* in Torrington Square, London (a plaque on the wall of University College, London, commemorates this event)

1815 Patents a type of ship's propeller, predating the invention of the modern screw propeller by 15 years

1816 Departs for Peru and Costa Rica

1827 Returns from Peru

1833 Dies on April 22 in Dartford, Kent, aged 62, having also invented the return-flue boiler, blast pipe, and coupled wheels, all of which were later developed by others, as well as a steam thresher and steam dredger used for dredging the River Thames

Opposite Richard Trevithick's circular railway in Torrington Square, near Euston, London (1808)
Right "Conjectural model" of Trevithick's locomotive *Catch Me Who Can*

The pneumatic brake, a hugely significant advance for the safety of railway trains, was invented by engineer and industrialist George Westinghouse, Jr., who was one of the world's most prolific inventors, with more than 400 patents to his name.

The inventor: George Westinghouse, Jr.

1846 Born on October 6 in Central Bridge, New York, and later educated in Schenectady, New York

1865 Receives his first U.S. patent on October 31 for a rotary steam engine. (Three months earlier his father had received a patent for a sawmill machine, sometimes erroneously credited to Jr.) December, leaves college and begins working in his father's farm machinery workshop

1867 Receives a patent on February 12 for a railway

AIR BRAKE

George Westinghouse, Jr. (U.S.) was the son of an inventive father who received some 30 patents of his own. Westinghouse Jr. worked at his father's farm machinery factory, G. Westinghouse & Co., before running away, aged 15, to fight for the Union Army in the Civil War. He returned to New York in 1865, at the insistence of his father, to attend college, and in October that year he received the first of his many patents. The college recommended that his mechanical aptitude would be better fulfilled in a workshop than a classroom, so in December he returned to his father's factory—but within two years he had a factory of his own.

After watching a railway crew struggling to put two derailed carriages back on the tracks, Westinghouse Jr. invented a machine to do just that. He received a patent for the rerailer in 1867 and established a factory the same year to manufacture his invention. Two years later he witnessed a head-on collision between two trains that led to his most important invention, the Westinghouse air brake. Because of the conventional braking system—which involved a brakeman in each carriage or wagon applying a separate brake—the trains had not been able to stop in time. Having read about compressed air being used for tunnel-drilling in Switzerland, where the drills operated up to 3,300 feet (1,000 miles) away from the compressor, Westinghouse applied the concept to braking trains. He invented a system that enabled the driver to apply the brakes in all the carriages or wagons at the same time, vastly increasing the safe speed of both freight trains and passenger trains.

Westinghouse Jr. is also famous as the champion of alternating current (AC) electricity in the face of fierce competition from Thomas Edison (U.S.), who favored direct current (DC). Westinghouse astutely bought patents for AC transformers and the AC induction motor invented by Nikola Tesla (Croatia–U.S.), and between 1891 and 1896 he established the first large-scale AC power station.

See also: Phonograph/Kinetoscope, page 158;
AC Induction motor, page 232

carriage rerailer and sets up his own factory to manufacture his invention

1869 Receives a patent on April 13 for his "steam power brake," later known as the Westinghouse air brake. (The patent is reissued in 1874 due to a technical error in the original.) Founds the Westinghouse Air Brake Co. on September 28

1876 Patents an automatic air brake system

1886 Founds the Westinghouse Electrical Co.

1888 Patents a quick-acting air brake system

1891 Patents a high-speed air brake system

1893 The Westinghouse air brake is made compulsory on all U.S. railways

1914 Dies of heart disease on March 12 in New York City, aged 67, having filed some 400 patents worldwide

including 361 U.S. patents. Of these patents, 20 relate to various aspects of air brakes, and others include railway signaling systems, natural gas pipelines and meters, a new steam turbine, the transmission and control of motive power, air suspension for road vehicles, and AC electrical power transmission

Below Investigators examine wreckage; air brakes would make derailments much less common
Right George Westinghouse, Jr. (1846-1914) patented the air brake and 100 other devices

John Loudon McAdam gave his name not only to the road surface that he invented in the 19th century, but also, indirectly, to Edgar Purnell Hooley's 20th-century invention, tar macadam, which has been known since 1903 as Tarmac.

MACADAMIZED ROADS

Roads in 18th-century Britain were notoriously bad—many of them were little better than mud tracks, and were completely impassable in wet weather, while those that had been paved with stones were equally bad because loose stones damaged the wheels of stage coaches and cart traffic. That began to change in the first quarter of the 19th century with the hundreds of miles of road built by engineer Thomas Telford (Scotland), who laid expensive foundations for his roads. But the biggest improvements were made by John Loudon McAdam (Scotland) from 1815 onward with a revolutionary new road surface that was impervious, hard-wearing, easily maintained, and, perhaps most important, cheaper than Telford's roads.

Instead of using foundations, McAdam pioneered the idea that roads could be built directly onto the subsoil if it was level and well drained, with a surface of compacted gravel and broken stone, cambered to improve drainage. In 1815 he was appointed surveyor to the Bristol Turnpike Trust and was able to put his ideas into practice. His roads soon became known as macadamized roads, and his system quickly proved such a success that he subsequently became surveyor to 18 turnpike trusts, and his system was adopted throughout Europe and the U.S.

Macadamized roads were superseded in the 20th century, but McAdam's name lives on in the name of their successor, Tarmac. McAdam's compacted stone surface was a great improvement on what had gone before, but it was still prone to develop ruts. In 1901 Edgar Purnell Hooley (England), a county surveyor, noticed that one particular stretch of road had no ruts—on investigation he was told that a barrel of tar had fallen off a wagon and that slag from a local ironworks had been used to cover the spillage. Using this as the basis for a new road surface, in 1902 Hooley patented a process for making what he called "tar macadam" and registered the trademark Tarmac the following year.

The inventor: John Loudon McAdam

1756 Born on September 21 in Ayr, Scotland

1770 Goes to New York and makes his fortune

1783 Returns to Scotland and begins experimenting with his new idea for a road surface

1815 Appointed surveyor to the Bristol Turnpike Trust and remakes the roads using the "macadamized" system

1825 Parliament votes to award McAdam £2,000 to reimburse his development costs

1827 Appointed surveyor general of roads in England

1836 Dies on November 26 in Moffat, Dumfriesshire, aged 80, having been surveyor to no fewer than 18 turnpike trusts and consultant to several more

1902 Edgar Purnell Hooley (England) receives a patent for: "Improvements in the means for and the method of"tarring" broken slag, macadam, and similar materials"

Below Tar spreader at work during dust-prevention experiments in Staines, Middlesex (May 1907) **Bottom** John Loudon McAdam

Did you know?

Edgar Hooley may have seemed gracious in perpetuating the name of his predecessor as part of the name of his invention, but the name of the company he formed to manufacture tar macadam was more self-promoting. He called it Tar Macadam (Purnell Hooley's Patent) Syndicate Ltd.—fortunately Sir Alfred Hickman (England), who bought the patent rights, shortened the name to Tarmac Ltd. in 1905.

After a hazardous drive home along an unlit road in 1933, Yorkshireman Percy Shaw came up with the idea for a gloriously simple invention that is still making night driving safer nearly 70 years later—Catseyes reflecting road studs.

CATSEYES

In the 21st century it is easy to forget how difficult it was to drive on unlit roads before the Second World War. Car headlights were inefficient, there was far less light pollution reflecting from the sky, and there were no reflecting road studs to mark out the course of the road. Like many other drivers, road contractor Percy Shaw (England) often used light reflected from tramlines built into the road to guide him, but by the 1930s many of Britain's tramlines were being removed. One night in 1933 Shaw was driving back to his home in Halifax when he saw his headlights reflected in the eyes of a cat, alerting him to the fact that he was veering off the road. He immediately thought that reflectors embedded in the road would make night driving much safer.

Shaw began testing various means of mounting reflective lenses, and said later: "I must have put them in the road 1,000 different ways. It was all a matter of trial and error." He eventually succeeded in designing a method that is basically unchanged today, and on March 15, 1935, he established a company, Reflecting Road Studs Ltd., to manufacture his invention. He filed a patent on May 30, 1935 (granted 1936), for a road stud incorporating two reflective lenses in a rubber housing mounted in a cast-iron base that could be set into the road. The concept was simple: If the studs were placed at intervals along the center of the road, they would light the way by reflecting car headlights—the clever part was that Catseyes were self-cleaning; when a car passed over them, the lenses were pushed into the rubber housing and would then pop up again, automatically cleaning themselves.

In 1937 the government tested Catseyes against ten other types of road stud, and after two years the Catseyes were the only ones still reflecting, so they were officially adopted just as war broke out. They proved invaluable during the wartime blackout, and have been an indispensable part of road safety ever since.

Left Reflecting roadstuds (c. 1990s)
Below Percy Shaw, inventor of Catseyes reflecting road studs, outside his factory in Boothtown, Halifax (September 1958)

The inventor: Percy Shaw

1890 Born in Halifax, Yorkshire

1934 Files a patent for a prototype reflecting road stud (granted 1935)

1935 Establishes Reflecting Road Studs Ltd. Files a patent for a self-cleaning reflecting road stud, later known as the Catseye (patent granted 1936, trademark registered 1938)

1947 Transportation Minister James Callaghan (later prime minister) orders millions of Catseyes for Britain's roads

1965 Shaw is awarded the OBE

1976 Dies, aged 86, having lived in the same house since the age of two

We have American journalist Carlton Magee to thank for this invention, the scourge of all motorists. Invented in 1932, the world's first parking meter was installed in Oklahoma City on July 16, 1935, and the first fine issued in August 1935.

PARKING METER

The idea of having no parking restrictions may sound like motoring utopia, but in 1930s Oklahoma it was a problem. Workers were arriving early and leaving their cars on the streets all day, preventing shoppers from parking and spending their money, so in 1933 the Chamber of Commerce set up a Businessman's Traffic Committee to discuss ways of solving the problem. Local newspaper editor Carlton C. Magee (U.S.) was appointed chairman; he suggested introducing parking meters, and the Oklahoma City Traffic Authority duly ordered 150 "Park-O-Meters" from the Dual Parking Meter Company (now POM Inc.), which just happened to be owned by Carlton C. Magee. (The name Dual Parking Meter Company was a reference to the dual purpose of the meters: controlling parking and raising revenue.) The first meters went into operation in Oklahoma City on July 16, 1935, and in August that year an Oklahoma churchman named North (U.S.) became the first driver in the world to be fined for overstaying at a parking meter.

Carl Magee, as he calls himself on his patent application, had filed his first patent for a parking meter in December 1932. The meter was little more than a box on a pole, but his improved patent of 1935 looked very similar to the parking meters still in use nearly 70 years later. This second application was filed on May 13, 1935, and granted just over three years later, on May 24, 1938.

Australia's first parking meters were introduced in Hobart, Tasmania, in 1955, and Magee's charming invention was introduced to Britain experimentally in 1956, followed by the first permanent installation by Westminster City Council in 1958, under whose auspices 625 meters went into operation in Mayfair, London, on July 10. Parking meters were soon followed by traffic wardens (aka meter maids), who first appeared in New York City and London in 1960 and who were celebrated by the Beatles in 1967 with their hit *Lovely Rita (meter maid)*.

Did you know?

Ernest Marples, who was appointed Minister of Transportation the year after parking meters were introduced to Britain, gave his name to what political satirists called Marples' Law: "If it moves, stop it; if it stops, fine it."

London's traffic wardens issued 344 fixed-penalty parking tickets on their first day in operation (September 19, 1960), the first being slapped on a Ford Popular belonging to a Dr. Thomas Creighton—who at the time was treating a patient suffering from a heart attack. The doctor was excused from paying his £2 fine after an outcry in the press.

Some modern parking meters operate using solar power.

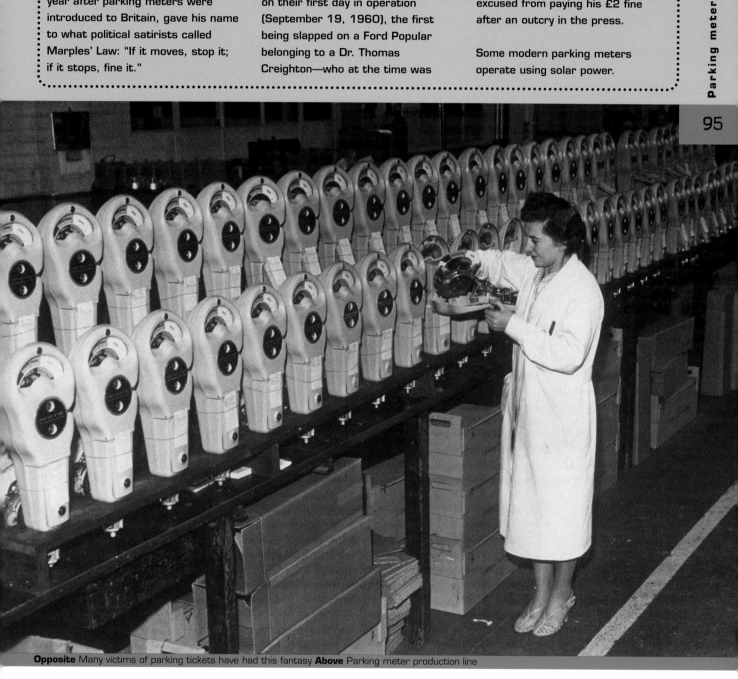

Opposite Many victims of parking tickets have had this fantasy **Above** Parking meter production line

Time line: POM Inc.

1932 Carlton C. Magee (U.S.) files the first patent for a parking meter and establishes the Dual Parking Meter Company

1935 Magee files a patent for an improved parking meter, the true forerunner of modern parking meters

1930s–1963 The company changes name to Hale-Magee Park-O-Meter Co., and becomes part of Rockwell International, making meters at factories in Oklahoma City and Tulsa

1963-64 Rockwell International moves manufacturing to Rusellville, Arkansas

1976 POM Inc. (Park-O-Meter Inc.) is incorporated to take over Rockwell International and its Arkansas factory

1992 POM markets a fully electronic parking meter, the "APM" (Advanced Parking Meter)

John Boyd Dunlop is often credited with inventing the pneumatic tire, but in fact what he had done was to reinvent it—the first patent for a pneumatic tire had been granted to Robert Thomson 42 years earlier.

The inventors: Robert Thomson & John Boyd Dunlop

1822 Robert Thomson is born in Stonehaven, Scotland, the son of a factory worker

1838 Thomson becomes assistant to a civil engineer and invents a system of firing explosives by electricity during the demolition of Dunbar Castle

1840 John Boyd Dunlop is born in Dreghorn, Scotland, the son of a farmer

1844 Thomson establishes his own business as a civil engineer

PNEUMATIC TIRE

The pneumatic tire was invented by civil engineer and inventor Robert Thomson (Scotland), who was granted a patent on December 10, 1845, for: "Improvement in carriage wheels which is also applicable to other rolling bodies." His tires were first demonstrated the following summer, to the amazement of *Mechanic's Magazine*, which reported the appearance of "a brougham [carriage] with silent wheels, so silent as to suggest a practical inconsistency of the most startling kind between the name and the quality of the thing," and went on to say that the tires could be "inflated with air to any degree of tightness desired." Thomson sold the patent rights to Messrs. Whitehurst & Co. in 1847, who began fitting them commercially, but the tires were too expensive and too impractical—with 70 bolts attaching them to the wheel—to be a success.

Pneumatic tires were then forgotten until they were further developed from 1887 to 1888 by John Boyd Dunlop (Scotland), who was working as a veterinary surgeon in Belfast. Noticing the deep ruts left in the garden by his son Johnnie's tricycle, Dunlop improvised some tires using lengths of garden hose filled with water. It was the family doctor, Sir John Fagan, who suggested filling the tires with air, and so the pneumatic tire was reborn. On July 23, 1888, Dunlop filed a patent for: "Improvement in Tires of Wheels for Bicycles, Tricycles, or other Road Cars," and by December he had secured a manufacturing agreement with bicycle manufacturers W. Edlin & Co. of Belfast.

The success of the tires encouraged Dunlop to set up his own manufactory, the Pneumatic Tyre Co., in 1899, but the following year *Sport and Play* magazine publicized the existence of Thomson's earlier patent, invalidating Dunlop's patent and allowing other manufacturers into the market. Despite the competition, the Pneumatic Tyre Co., which later became the Dunlop Rubber Company, continued to be a success, ensuring that Dunlop's name is still associated with tires today.

Above Johnny Dunlop (aged about ten) riding his tricycle with rubber tires (1888) **Opposite** 1930s Dunlop advertisement

Did you know?
Dunlop's original tires were fixed to the wheel by sticking them to the rim, making them difficult to change. It was not long before the situation was remedied—in 1890 Charles Kingston Welch (Britain) invented the "wired" tire still in use today.

The first person to fit pneumatic tires to a motor car was Edouard Michelin (France), in 1895.

1845 Thomson patents the first pneumatic tire

1849 Thomson patents a fountain pen

1852 Thomson works in Java as the agent for a manufacturer of sugar-refining machinery, where he invents a portable crane and a hydraulic dock

1857 Dunlop qualifies as a veterinary surgeon

1862 Thomson retires to Edinburgh, Scotland

1867 Thomson patents a steam traction road haulage vehicle known as the Thomson Steamer (first produced commercially in 1872 and

later exported around the world). Dunlop moves his veterinary practice to Belfast

1873 Thomson dies

1887 Dunlop reinvents the pneumatic tire

1888 Dunlop files a patent for his pneumatic tire

1890 The existence of Thomson's patent becomes public knowledge

1921 Dunlop dies in Dublin

Amphibious vehicles are nothing new, but one that is capable of 154 mph (257kph) on land and 48 mph (80 kph) on water surely is—the Land Shark, with its revolutionary centrifugal turbine propulsion system, was invented in 1997 by David Baker.

LAND SHARK

Looking as though it has just driven out of a James Bond movie, the Land Shark is a revolutionary three-wheeled amphibious vehicle invented by David Baker (England) that will have many potential uses in leisure, firefighting, rescue operations, policing, and the military—once a full-size model is built. On entering the water the front mudguards of the Land Shark swing out and down through 180 degrees to sit beneath the wheels and act as hydroplanes, while the single rear wheel acts as a turbine pump. One side of the rear wheel draws the water into a centrifugal turbine; the water is then pushed through to a pressure collector on the other side and forced out of the vehicle via a rear-facing nozzle. It was the realization that such a wheel/pump could work that inspired Baker to invent the Land Shark after he noticed a motorbike with an aluminum rear wheel: "The centre of the spokes resembled a propeller. If you had dropped the back wheel of the motorcycle into the water, it would have pumped water straight through the middle of the wheel."

Inspired by this thought, Baker went on to invent a lightweight, three-seater vehicle that has been described as a cross between a car, a motorbike and sidecar, a jet boat, and a hydrofoil. He filed a patent in 1997 and managed to attract the interest of Britain's Defense Evaluation Research Agency (DERA), which joined the project late that year. Land Shark Ltd. was incorporated in 1998 to bring the invention to fruition, and that autumn Lotus Engineering Ltd. (England) joined the Land Shark Project as consultants, later agreeing to design and build the first prototype.

Baker asserts that the biggest advantage of the Land Shark over earlier amphibious vehicles is that: "Most have been adapted from standard road car designs and this has always added to the cost and complexity of the finished vehicle—due to the addition of a second drive system for use in water." The Land Shark, on the other hand, has a single system that drives the ingenious rear wheel, which doubles as a centrifugal turbine.

Did you know?

The outer skin of the Land Shark will be made from PET (polyethylene terephthalate)—in other words, recycled drink bottles.

David Baker brought his invention to the world's attention through Inventorlink, an organization that liaises between inventors and potential manufacturers. Inventorlink bills itself as "Innovation's link with industry."

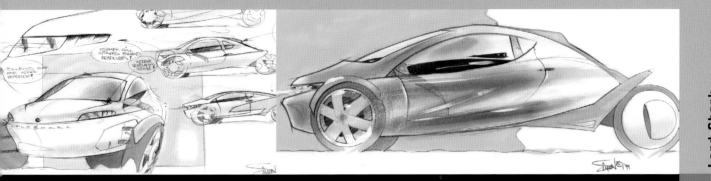

Opposite top and Below Computer graphics of how the Land Shark will look in the water and on land **Above center and right** Concept drawings of the Land Shark by Simon Grand

The inventor: David Baker

1966 Born on January 9 in London and later attends Bacon Secondary School, London

1983 Becomes a stone carver with English Heritage

1984–87 Studies at Kennington School of Art while apprenticed to English Heritage

1997 Invents and files a patent for the Land Shark (granted May 14, 2002). Helps sculpt a full-size fiberglass Indian elephant for Harrods department store, London

1998 Forms Land Shark Ltd. on August 19 to build his invention

2000 Works on the Wellington Arch in London, carving replacement sections of the capitals and a lion's mask for the main cornice

2003 Becomes mason foreman at the House of Commons restoration in London

Legend has it that the Patent Office did not know whether to classify the hovercraft, invented by Christopher Cockerell in 1955, as a boat or an aircraft, because it did not float on the water, but neither did it fly through the air.

HOVERCRAFT

Christopher Cockerell (England) was an electrical engineer who bought himself a boatyard in order to pursue a personal interest in boat-building. As a scientist, he was interested in increasing the speed and efficiency of boats by reducing the hydrodynamic drag (i.e., friction from the water), and he began to investigate the idea of using a cushion of air so that the boat was not in contact with the water at all.

One of Cockerell's early experiments was to put two tins inside each other (reputedly a coffee tin and a cat food tin), the larger with both ends removed. He used the nozzle of a vacuum cleaner (with the motor reversed) to blow air into the narrow gap between the walls of the two tins, and found that the pressure of the air coming out of the tins was three times greater than if the small tin was removed—a concept known as the "annular jet." Cockerell applied this principle to design a hovercraft, which had several annular jets around the base of the craft, pointing downward and inward to provide lift.

The first hovercraft was SRN1, built by Saunders-Roe, a British aircraft manufacturer that had previously built flying boats and therefore had maritime and aeronautical expertise. SRN1 was launched on May 30, 1959, and successfully underwent trials off Cowes, Isle of Wight, before making the first hovercraft crossing of the English Channel—aptly, the crossing was made on July 25, the 50th anniversary of Frenchman Louis Blèriot's historic first cross-Channel flight. However, there were limitations to Cockerell's design, which could only clear relatively small waves or obstacles despite a clever "closed vortex" system. This problem was solved by the addition of a flexible skirt, an invention sometimes credited to Cockerell and sometimes to C.H. Latimer-Needham (England). The skirt worked by trapping the air cushion underneath the hovercraft, giving much greater clearance and doubling the payload for a given power output.

The inventor: Sir Christopher Cockerell

1910 Born Christopher Sydney Cockerell in Cambridge and educated at Gresham's School, Holt, England

1931 Graduates from Cambridge University with a degree in engineering

1931–33 Works for engineering firm of W.H. Allen & Sons

1933–35 Returns to Cambridge University to study radio

1935–50 Works for Marconi Wireless Co. developing VHS transmitters and direction finders. During the Second World War he develops radar, airborne navigation systems, and communications systems, filing no fewer than 36 patents on behalf of Marconi. Among the most significant of his patents are Frequency Division (1935), Linearization of a Transmitter by Feedback

Did you know?

The first hovercraft passenger service began on July 20, 1962, between Wallasey, Cheshire in England and Flint, Rhyl in North Wales. The first scheduled passenger service was inaugurated in 1968 between Dover, England, and Boulogne, France.

Cockerell sold his patent rights to the British government in 1971, but it is said that he did not even recover his development costs. Cockerell said: "I had to take what I was offered. It's no big figure when you look at what was given to the inventors of radar and jet engine, but I've been living on air for some time now." The government later sued the U.S. Army for patent infringement and won six million dollars in damages.

Opposite top Cockerell's original model of a hovercraft (c. 1950s) **Below** Sir Christopher Cockerell watches the British Hovercraft Corporation's Princess Margaret roar up the River Thames (May 1979)

(1937), and Pulse Differentiation (1938)

1950 Leaves Marconi and moves to Suffolk, where he buys a boatyard

1955 Files a patent on December 12 for: "Improvements in or relating to vehicles for travelling over land and/or water"—the first hovercraft (granted 1960)

1956 Devises a closed vortex system to make the air cushion more efficient. Demonstrates a model of the hovercraft to the government, which at first classifies the invention as secret

1957 Develops the hovercraft further under the auspices of the National Research and Development Corporation (NRDC)

1959 The hovercraft is declassified. NRDC forms Hovercraft Development Ltd., with Cockerell as a director, to continue development. The first hovercraft is launched

1965 Granted a patent for: "improvements relating to vehicles for travelling along a prepared track," describing an air-cushioned vehicle

1969 Receives a knighthood

1971 Sells his patent rights to the British government

1970s Develops his interest in wave power

1974 Becomes chairman of Wavepower Ltd. Made an Honorary Fellow of his alma mater, Peterhouse, Cambridge

1999 Dies on June 1 in Hythe, Hampshire, aged 88

The invention of a means of buoying vessels over shoals was unremarkable except for the identity of the inventor—Abraham Lincoln, who is not the only U.S. President to have been an inventor but is the only one to have filed a patent.

BUOYING VESSELS

On May 22, 1849, Abraham Lincoln (U.S.) was granted U.S. Patent No. 6,649 for a: "Manner of buoying vessels." The invention was never put into practice, but the patent is significant because it is the only one ever issued to a future U.S. President. The patent states: "To all whom it may concern: Be it known that I, Abraham Lincoln, of Springfield, in the County of Sangamon, in the State of Illinois, have invented a new and improved manner of combining adjustable buoyant air chambers with a steamer or other vessel for the purpose of enabling their draft of water to be readily lessened to enable them to pass over bars, or through shallow water, without discharging their cargoes."

Lincoln had practical experience to draw on for his invention, having made two trips down the Mississippi as a hired hand on flatboats carrying cargo to New Orleans. On the second trip Lincoln's boat went aground on a mill dam, and on another occasion, years later, a boat on which he was a passenger ran aground on a sand bar. Lincoln watched as the captain used empty barrels to buoy up the boat and float it free of the sand bar; this gave him the inspiration for his invention, which involved attaching inflatable chambers to the sides of boats, an idea that he predicted would revolutionize steam navigation. This was not the case, but, nonetheless, the model he built to support his patent application is kept in the Smithsonian Institution. Lincoln was a great supporter of the patent system, and once said that the three most important developments in world history were the discovery of America, the invention of printing, and the introduction of patent protection.

Lincoln was not the first inventive President: George Washington invented a plow and a wine coaster, and Thomas Jefferson a swivel chair, a folding buggy, a stool, a plowshare, a spherical sundial, and a writing desk, but neither applied for patents—both, however, signed U.S. Patent No. 1, in 1790, in their roles as President and secretary of state respectively.

The inventor: Abraham Lincoln

1809 Born on February 12 in a log cabin near what is now Hodgenville, Kentucky

1828 Makes his first trip down the Mississippi River on a flatboat

1830 The family moves to Illinois, where Lincoln works as a shop clerk

1831 Makes his second trip down the Mississippi on a flatboat

1834 Elected to the Illinois General Assembly

1836 Begins to practice law

1846 Elected to the House of Representatives

1849 Files his patent application on March 10 (granted May 22)

1860 Elected President of the U.S. on November 6

1863 Delivers his famous Gettysburg Address

1864 Elected for a second term as President on November 8

1865 Shot by John Wilkes Booth at Ford's Theatre, Washington D.C., on April 14 and dies the following morning, aged 56 See also: Pullman cars, page 116

Did you know?

In 1859 Abraham Lincoln said the patent system "secured to the inventor, for a limited time, the exclusive use of his invention; and thereby added the fuel of interest to the fire of genius, in the discovery and production of new and useful things."

Another American statesman who was also an inventor was founding father Benjamin Franklin. He invented (among other things) the Franklin stove and the first bifocal spectacles, and is famous for his pioneering electrical experiments.

Other patent-holders who are better known for things other than their inventions include Danny Kaye, Zeppo Marx, Lee Trevino, Mark Twain, Hedy Lamarr, Harry Houdini, and Jamie Lee Curtis.

Opposite top Abraham Lincoln, as depicted in the Lincoln Memorial, Washington, DC **Below** Riverboats on the Mississippi River (1870)

As with aircraft, the idea of submarines had existed a long time before anyone managed to design one that worked. The first realistic design dates from 1578 and the first working submarine, a wooden vessel clad in leather, from c. 1624.

SUBMARINE

The first serious design for a submarine was proposed by mathematician William Bourne (Britain) in 1578. Bourne never tested his design, but just over 40 years later Cornelius Drebbel (Netherlands–Britain) built a submarine of remarkably similar design and tested it in the River Thames, England, c. 1624. It had a wooden frame enclosed within a leather skin, and was propelled by 12 oars protruding through sealed ports in the leather casing. Drebbel's submarine was reported to have remained submerged for two hours, leading to speculation that Drebbel had found a means of producing oxygen, some 150 years before its official discovery. Despite the success of the demonstration, the British Admiralty did not adopt Drebbel's submarine, and it was left to later inventors to develop the concept further.

The first submarine to be used in warfare was invented by David Bushnell (U.S.) in 1775 and was known as the *American Turtle*. It was a wooden vessel driven by two hand-powered screw propellers, and first saw action in 1776 during the Revolutionary War, though it failed to destroy its intended target. The first submarine to actually destroy an enemy ship was the *Hunley*, designed by H.L. Hunley (U.S.) for the Confederate Army in 1862, which sank the Union warship *Housatonic* in 1864, but was itself destroyed in the process, killing all the crew.

Despite these and many other early submarines, J.P. Holland (Ireland–U.S.) is acknowledged to be the inventor of the modern submarine. Holland's vessel was the first to have all the requirements of a modern submarine: a circular transverse section, ballast tanks, depth control, horizontal rudders, and a power supply that does not rely on air—his submarine used batteries when submerged and internal combustion while on the surface. Holland provided the U.S. Navy with its first submarine in 1900, was granted a patent for his "submarine boat" in 1902, and went on to build submarines for Great Britain, the Netherlands, Russia, and Japan.

See also: Leonardo's inventions, page 84

Below John Holland standing in the turret of one of his early submarines (c. 1898)
Right The Beatles' immortal yellow submarine

The inventors: Cornelius Drebbel & John Philip Holland

c. 1572 Drebbel is born Cornelius Jacobszoon Drebbel (sometimes recorded as van Drebbel) in Alkmaar, Netherlands, and later apprenticed to a painter, engraver, and alchemist in Haarlem

1604 Drebbel immigrates to England

c. 1624 Drebbel demonstrates the world's first submarine

1633 Drebbel dies in England, having invented a clock driven by changes in atmospheric pressure, a method for the manufacture of sulfuric acid, and a thermostat for furnaces

1840 John Philip Holland is born on February 29 in Liscannor, Ireland, and is educated in Liscannor, Ennistymon, and Limerick

1858–72 Holland works as a teacher in Ireland

c. 1870 As a supporter of Irish independence, Holland conceives the idea of a submarine to be used against the British

1873 Holland immigrates to the U.S.

1875 Holland's proposal for a submarine is rejected by the U.S. Navy

1881 Holland is funded by the Fenian Society to build his first submarine, the *Fenian Ram*

1895 The U.S. Navy awards Holland a contract to build a submarine. Most of Holland's designs are overruled by a U.S. Navy admiral and the resulting vessel, *Plunger*, is abandoned as a failure

1898 Holland builds his own submarine, *Holland*, to his own designs. It is a resounding success and in 1900 it becomes the U.S. Navy's first submarine

1902 Holland is granted a patent for his "submarine boat"

1904 Holland retires and spends his time conducting aeronautical experiments

1914 Holland dies on August 12 in Newark, New Jersey, aged 74

The history of flight began long before the Wright brothers: The Montgolfiers started flying balloons in 1783, and the first heavier-than-air craft to carry a human being—a glider—was invented by Yorkshireman George Cayley in 1849.

The inventor: Sir George Cayley

1773 Born on December 27 in Scarborough, North Yorkshire, and is later taught by scientist and mechanic George Walker (England)

1799 Designs the first fixed-wing aircraft since Leonardo da Vinci's glider of 1490 *See*

also: Leonardo's inventions, page 84

c. 1803 Designs a form of caterpillar track a century before David Roberts (England) invents the first "crawler tractor"

GLIDER

For many years, the significance of the glider invented by George Cayley (England) was overlooked, but he is now universally acclaimed as the father of flight. His pioneering work was recognized by the Wright brothers (U.S.), who acknowledged that they had based the design of their *Wright Flyer* on Cayley's glider. The Wright brothers are famous for making what is generally accepted to be the first sustained, manned, controlled, powered flight in a heavier-than-air craft in 1903, but more than 50 years earlier, in 1849, Cayley had been the first to put a human being in the air in a heavier-than-air craft.

Cayley's interest in aviation began when he read about the model helicopters made in 1784 by Launoy and Bienvenu (both France). Inspired by the idea that something heavier than air could fly, in 1799 Cayley produced his first design for a fixed-wing aircraft, built a scale model in 1804, and flew a full-size model in 1809. He went on to build thousands of scale model aircraft, which he is said to have tested on the staircase of the family seat at Brompton Hall, Yorkshire, much to the annoyance of his wife. He returned his attention to full-scale flight in 1849, when he built a three-winged glider that succeeded in carrying a ten-year-old boy on the first (tethered) manned flight by a heavier-than-air craft.

Then, in 1853, came the flight that secured his fame as the father of flight. On September 25, 1852, Cayley published designs in *Mechanics' Magazine* for a single-wing glider that could be controlled by its pilot and was capable of carrying an adult. The following June Cayley persuaded his coachman, John Appleby (England), to fly the glider across a valley in the grounds of Brompton Hall, making the first manned free flight of a fixed-wing aircraft and the first manned free flight of a heavier-than-air craft. Little realizing the importance of his place in the history of flight, Appleby said on landing: "Please, Sir George, I wish to give notice. I was hired to drive, not fly."

1808 Designs the first tension-spoked wheel, which he uses on his aircraft undercarriages

1809 Builds and flies his first full-scale model aircraft, which is capable of lifting a person off the ground

c. 1830 Witnesses a crash on the new Liverpool–Manchester

Railway, which prompts him to invent a passenger seat belt

1839 Helps to found the Regent Street Polytechnic in London (eventually part of the University of Westminster)

1849 Builds a three-wing glider that carries a 10-year-old boy in the first manned flight

1852 Publishes designs for a single-wing glider capable of carrying an adult

1853 His coachman makes the first manned free flight of a fixed-wing aircraft. Makes a model aircraft powered by a stretched rubber band (proving he was considering the concept of powered flight)

1857 Dies on December 15 in Brompton, Yorkshire, aged 83, having also invented a gunpowder engine (an early form of internal combustion engine), a cowcatcher for trains and a prosthetic hand for the son of one of his tenants

Glider

Opposite Sir George Cayley, English aviation pioneer (c. 1850) **Below** Sir Richard Branson plays the part of the coachman in a celebration of the 150th anniversary of the first manned free flight of a fixed-wing aircraft (Brompton, 2003)

107

Did you know?

In June 2003 Sir Richard Branson (England) played the part of Cayley's coachman, John Appleby, in a reconstruction to celebrate the 150th anniversary of the historic glider flight. Branson funded BAe Systems and the Royal Aeronautical Society to build a replica of the glider, which he flew himself. Afterwards, Branson's reaction was more positive than Appleby's had been: "That was exhilarating, magnificent. I can retire knowing that I can fly."

The jet engine was invented by aeronautical engineer Frank Whittle, patented in 1930 and bench-tested in 1937. But the first jet engine into the air was invented independently by the German engineer Hans von Ohain, and first flew in August 1939.

JET ENGINE

Above Frank Whittle signs copies of his book *Jet, The Story of a Pioneer* for waiting schoolboys at an exhibition in London (January 1954)

The idea of jet propulsion is fairly simple, being the embodiment of Newton's Third Law of Motion that for every action there is an equal and opposite reaction: In essence, a jet of gas or liquid leaving the back of a vehicle at force will propel the vehicle forward. Scientists had experimented with the idea for centuries, and a steam-driven jet engine was used to power a boat as early as 1781, yet Frank Whittle's invention was so revolutionary that it took several years to gain official acceptance.

In 1928 Whittle (England) wrote a paper suggesting the use of jet propulsion or gas turbines as an alternative to internal combustion for powering aircraft. The following year he made the breakthrough of combining the two ideas, using a gas turbine to power a jet engine, and on January 16, 1930, he filed a patent for his turbojet engine, which was granted in 1931. (In 1941 he filed the first U.S. patent for a jet engine, which was granted on July 16, 1946.) Lack of money or encouragement meant that Whittle did not develop his idea any further until 1936. He formed a company called Power Jets, built a prototype, and on April 12, 1937, demonstrated the world's first turbojet engine. Four years later, Britain's first jet fighter, the Gloster-Whittle E.28/39, took to the air.

But it was not the *world's* first jet fighter: Hans von Ohain (Germany) had been working on jet engines in Germany since 1933, and patented a centrifugal-type turbojet engine in 1934. In June 1938 von Ohain's engine was flight-tested beneath a modified Heinkel He 118, and on August 24, 1939, a He 178 made the first flight by a turbojet-powered aircraft, followed by an official demonstration three days later. Whittle and von Ohain both later immigrated to America, where in 1987 they were jointly awarded the National Air and Space Museum Trophy in recognition of their work.

Did you know?

The world's first regular scheduled jet airline service (between London and Johannesburg, South Africa) was inaugurated by British Overseas Airway Corporation on May 2, 1952, using the De Havilland *Comet.*

The inventor: Frank Whittle

1907 Born on June 1 in Coventry, England, the eldest son of a mechanic who owned and ran a factory manufacturing valves and piston-rings

1918 Attends Leamington College

1923 Becomes an aircraft apprentice at RAF Cranwell, England

1926 Selected for officer and pilot training

1928 In a thesis for the RAF college, Whittle suggests the use of rocket propulsion and of gas turbines—two ideas that will later be combined in his turbojet engine. Assigned as test pilot to the Marine Aircraft Experimental Establishment. Later continues studying at the Central Flying School at Wittering, where he

The jet-powered Gloster *Meteor* was the only Allied jet aircraft to participate in the Second World War. It was descended from the Gloster-Whittle E.28/39, which had been conceived and built in just 15 months. The E.28/39's first test flight took place at RAF Cranwell, England, on May 15, 1941, piloted by Flight Lt. Gerry Sayer; the *Meteor* saw active service on July 27, 1944.

The Messerschmitt Me 262A was the first jet aircraft to be used in active warfare. It engaged a conventional RAF *Mosquito* on July 25, 1944—the RAF pilot escaped into clouds. The first jet-to-jet aerial combat took place in 1950 during the Korean War, when a U.S.A.F. Lockheed F-80 destroyed a Soviet MiG-15.

Above An F-4 Phantom brings up its landing gear as it takes off from a military airfield

combines his earlier ideas to invent the turbojet engine

1930 Files a patent for the idea of the jet engine (granted 1931), but the idea is ignored by the Air Ministry

1934 Sent to Cambridge University by the RAF to study engineering

1935 His patent protection lapses due to lack of money. Two ex-RAF officers at

Cambridge encourage him to pursue the idea nonetheless

1936 Forms Power Jets

1939 Power Jets is awarded an Air Ministry contract to develop a jet engine; Gloster Aircraft Co. is awarded the contract to build a plane to go with it

1941 Britain's first jet fighter, the Gloster-Whittle E.28/39, makes its maiden flight

1944 Power Jets is nationalized

1946 Whittle is awarded the Daniel Guggenheim Medal for the development of the jet engine

1947 Elected a Fellow of The Royal Society

1948 Invalided out of the RAF at the rank of air commodore. Knighted by King George VI

1953 Publishes *Jet, The Story of a Pioneer*

1976 Immigrates to the U.S., where he becomes a research professor at the U.S. Naval Academy in Maryland

1987 He and Hans von Ohain (Germany) are jointly awarded the U.S. National Air and Space Museum Trophy

1996 Dies on August 8 in Maryland aged 89

Radar now has an important role in keeping the seas and the skies safe for traffic, but it began life with the opposite intention when its inventor, Robert Watson-Watt, was asked to investigate the possibility of building a "death ray."

RADAR

Robert Watson-Watt (Scotland) is generally credited as the inventor of radar, but, as with so many inventions, there were precursors and false starts before Watson-Watt invented the first practical system. As early as 1904, Christian H. Ismeyer (Germany) patented a system of "employing a continuous radio wave to detect objects" as a collision-warning system for ships. A patent filed in 1926 by John Logie Baird (Scotland) describes "a method of viewing an object, by projecting upon it electromagnetic waves of short wavelength," which constitutes a form of radio detection, and in 1933 Rudolf Kühnold (Germany) developed radio detection equipment and demonstrated it in Kiel Harbor the following year.

However, a crucial function of radar is rangefinding, and the first equipment to enable detection and ranging was invented in 1935 by Watson-Watt, who had previously researched the use of cathode ray direction finders to detect thunderstorms (patented in 1926). In January 1935, Watson-Watt was asked by the Committee for the Scientific Survey of Air Defence to investigate whether radio waves could be used to destroy enemy aircraft. On February 12 he replied that radio waves would be insufficient to actually destroy an aircraft, but that planes could be detected by bouncing radio waves off them, measuring the delay before the echo returned, and thereby calculating the direction and distance of the aircraft. Two weeks later, on February 26, he demonstrated radar by plotting the course of a Heyford bomber from eight miles away, and on September 17 he filed a patent for the first practicable radar system.

Until 1943 radar was known in Britain as RDF, or radio direction finding. Meanwhile, it was developed independently in Germany and in the U.S., where U.S. Navy Cmdr. S.M. Tucker coined the phrase radar from radio detection and ranging.

See also: Television, page 152

Did you know?
Professor P.M.S. Blackett (Britain), a member of the Committee for the Scientific Survey of Air Defence, which commissioned Watson-Watt to investigate the "death ray," said that without radar "the Battle of Britain in 1940—a near thing at best—might have been lost, with incalculable historic consequences."

The inventor: Robert Watson-Watt

1892 Born Robert Alexander Watson Watt on April 13 in Brechin, Scotland (only later does he hyphenate his surname)

c. 1910 Educated at Dundee University and the University of St. Andrews

c. 1915–40 Works in the Meteorological Office, at the Dept. of Scientific & Industrial Research, and at the National Physical Laboratory and is appointed superintendent of the Radio Research Laboratory

1935 Presents a paper entitled "Detection and Location of Aircraft by Radio Methods." Demonstrates a working radar system. Files a patent for: "Improvements in or relating to wireless systems" (not granted until 1947 on the grounds of secrecy)

1940 Appointed scientific adviser to the Air Ministry

1941 Elected a Fellow of The Royal Society

1942 Receives a knighthood

1958 Publishes *Three Steps to Victory*

1973 Dies on December 5 in Inverness, aged 81

Opposite top Robert Watson-Watt experiments with a kite and radio transmitter (1931) **Below** Crew members use radar screens to monitor air operations from the combat information center of the aircraft carrier USS *Kitty Hawk* while at sea in the Atlantic Ocean (1991)

More popularly known as the black box, the flight data recorder was invented by David Warren during the 1950s. Thanks to this invention it has been possible to determine the cause of many aviation disasters and to prevent them from being repeated.

FLIGHT DATA RECORDER

"Orange box" does not have the same ring to it as "black box," but it would be a more accurate name, because flight data recorders are actually housed in high-visibility orange boxes to make them easier to find in the aftermath of an accident. The idea of recording flight data for use in such an event seems very obvious now, but when David Warren (Australia) came up with the idea during the 1950s he had difficulty convincing anyone that it was a worthwhile invention.

In 1953, while working as a research scientist for Australia's Aeronautical Research Laboratories, Warren was involved in the accident investigations following the crash of the world's first jet airliner, the de Havilland *Comet*. He realized that it would be very useful to know what had been happening in the cockpit at the time of the crash and, remembering a miniature tape recorder he had seen at a trade fair, he began to ponder the possibilities of a machine that would record the instrument readings and the voice of the pilot.

Warren wrote a proposal outlining the details of his idea and sent it to aviation authorities in several countries, but he could not interest anyone in it. Undeterred, he built a prototype that could record four hours of speech as well as information such as the speed, height, and direction of an aircraft, with the data stored on a steel wire rather than a tape, making it far less likely to be damaged by fire. Eventually, in 1958, Warren had the opportunity to demonstrate his prototype to former Air Vice Marshal Sir Robert Hardingham (Britain), who was visiting the Aeronautical Research Laboratories. Realizing its potential, Hardingham immediately promoted the idea in Britain, where it was approved by the Ministry of Aviation and use of the recorder in aircraft was later made compulsory. Aptly, the first country to make black boxes compulsory was Australia, after the judge investigating a 1960 air crash in Queensland recommended that all aircraft should be equipped with flight recorders.

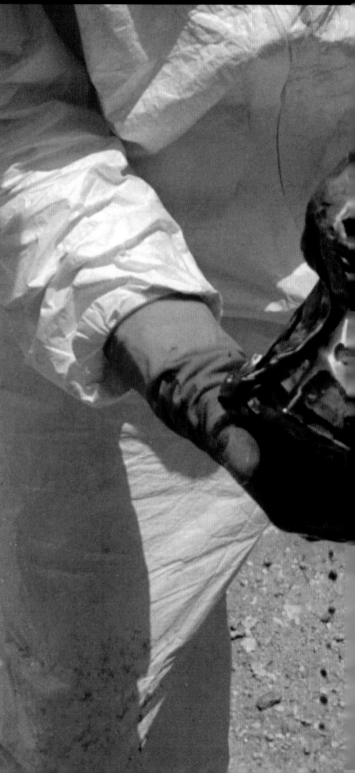

Left A flight data recorder being recovered after an air crash in Miami, Florida (1996) **Below** Allied Signals's new digital version of the black box (1999)

113

The inventor: David Warren

1925 Born on a mission station in Groote Eylandt, Northern Australia

1934 His father is killed in one of Australia's earliest air crashes

1944–46 Warren teaches mathematics and chemistry at Geelong Grammar School, Victoria

1947–48 Lectures in chemistry at the University of Sydney

1948–51 Employed as scientific officer at the Woomera Rocket Range

1952 Becomes a research scientist at the Aeronautical Research Laboratories, Melbourne (now part of the Defence Science and Technology Organisation), where he works until 1983

1953 Working for Aeronautical Research Laboratories, he is involved in the investigation of the crash of a de Havilland *Comet* and comes up with the idea of a flight recorder

1958 Demonstrates a prototype flight recorder to Sir Robert Hardingham

1981–82 Acts as scientific adviser (Energy) to the Victorian State Parliament

1999 Awarded the Australian Institute of Energy Medal

2000 Awarded the Hartnett Medal of The Royal Society of the Arts

2001 Awarded the Lawrence Hargrave Award of the Aeronautical Society

2002 Appointed Officer in the General Division of the Order of Australia in the Australia Day honors list, for "service to the aviation industry, particularly through the early conceptual work and prototype development of the black box flight data recorder"

FLIGHT RECORD DO NOT

"Jubilee Clip" is often used erroneously to describe any worm-drive hose clip, but the name is actually a registered trademark of L. Robinson & Co. The Jubilee Clip was the first such clip, and was invented by Cmdr. Lumley Robinson in 1920.

Opposite A *Spitfire* from 303 "Kozciuszko" Squadron (Poland) patrols the skies during the Second World War (c. 1942) **Below** Advertisement celebrating the golden jubilee of L. Robinson & Co., showing some of the many applications of Jubilee Clips (1971)

JUBILEE CLIP

Most people will have come across the Jubilee Clip or its imitators on their garden hose, but in fact this ingenious clip was invented as a precision engineering device for use in engines and machinery, particularly in the aviation and automotive industries. Jubilee Clips are now used in everything from domestic appliances to fighter planes and Formula One racing cars.

The Jubilee Clip was invented in 1920 by Cmdr. Lumley Robinson (England), and comprises a grooved metal collar with a worm drive to tighten the collar around the hose, securing it firmly to a tube or pipe. Robinson served in the Royal Navy during the First World War, and he perceived the need for a hose clip while working in the engine room of his ship. Three years after the war ended, he invented the Jubilee Clip and founded L. Robinson & Co. to manufacture it; when he died in 1939, on the eve of the next world war, his wife, known to all as Mrs. E.B., took control of the company. Wartime demand for Jubilee Clips was huge, particularly for use in military aircraft, and Mrs. Lumley-Robinson (who had hyphenated her surname in order to preserve her late husband's given name) rose to the challenge, producing Jubilee Clips at a rate of more than one million a month.

L. Robinson & Co. (Gillingham) Ltd. remains a family firm; the founder's son, John, was managing director from 1948 until 1988, when his son-in-law, John C.B. Jennings (England), took over. Jennings puts the success of the Jubilee Clip down to good old-fashioned values: "Our patent actually ran out in 1948 and since then we've had numerous competitors. Where we profit over them is that we adhere to the Commander's original idea that quality comes first." So the secret of success is simple, but the secret of the name, now registered as a trademark in 130 countries, remains a mystery—there was no royal jubilee in 1921, and the commander never told anyone the reason why he had chosen the name Jubilee for his clip.

Did you know?

After hearing a radio appeal in 1941 by the Air Ministry for funds, Mrs. E.B. Lumley-Robinson donated £5,000 to buy a *Spitfire* for the RAF.

L. Robinson & Co. (Gillingham) Ltd. time line

1877 Cmdr. Lumley Robinson born in Leeds, England

1914–18 Serves in the Royal Navy and conceives the idea of a hose clip

1920 Invents and files a patent for his hose clip (granted 1921)

1921 Founds L. Robinson & Co. to manufacture his invention

1928 Files a patent on February 13 for improvements to the Jubilee Clip (granted July 5)

1939 Dies in September and his wife takes over the company

1948 John S. Lumley-Robinson (son of the inventor) becomes managing director

1963 The Jubilee Clip is granted BS3628, the first British Standard for hose clips/clamps

1988 John S. Lumley-Robinson's son-in-law, John C.B. Jennings, becomes managing director

1982 Mrs. E.B. Lumley-Robinson retires from the company at the age of 97

1985 Mrs. E.B. Lumley-Robinson dies

1998 John S. Lumley-Robinson dies and is succeeded as chairman by his widow, Mrs. G. Lumley-Robinson, who is the last person at the company to have been employed by Commander Robinson

2003 I.P.J. Jennings and Miss E.M. Jennings join the board of directors to become the fourth generation of the family involved in the company. (Mrs. M.G. Jennings née Lumley-Robinson is also a director.) The Jubilee Clip is awarded ISO 9001:2000, the latest International Quality Assurance Standard

AND ALSO...

Pullman cars

The railway sleeping carriage was patented by former cabinetmaker George Pullman and his associate Ben Field (both U.S.) in 1865. Their invention was given a huge publicity boost when a Pullman carriage was used to take the body of President Lincoln home to Illinois from Washington, D.C., that same year. Pullman and Field formed the Pullman Palace Car Co. in 1867 to manufacture and operate their carriages, and in 1869 they took out further patents for a dining carriage and a hotel carriage, since which time Pullman has become a by-word for luxury on the railways.

See also: Buoying vessels, page 102; Railway locomotive, page 86

Bowden cable

The Bowden cable is used in everything from the humble bicycle brake cable to aeronautical engineering. It comprises a cable in the form of a flexible metal spiral covered with a plastic outer sheath, within which runs an inner cable that does the pulling. Usually a wire can exert force only if it is stretched in a straight line, but the outer spiral of the Bowden cable enables the inner wire to pull even when curved to fit the frame of, for instance, a bicycle. The cable is named after its inventor, Ernest Mannington Bowden (England), who was no relation to Frank Bowden of Raleigh fame. In 1896 Bowden was granted a patent for a: "New and improved mechanism for the transmission of power," and over the next two years he was granted two further patents relating to the use of his cable for bicycle brakes.

Hub gears

Today the name Sturmey-Archer is almost synonymous with bicycle hub gears. The story begins in 1887, when Frank Bowden (England) decided to produce a practical system of bicycle gears after being forced to dismount frequently on a cycling trip to the Pyrenees. The following year Bowden acquired a financial stake in a bicycle workshop in Raleigh Street, Nottingham (later Raleigh Industries Ltd.), and in 1902 he found that schoolmaster Henry Sturmey and engineer James Archer (both England) had each independently invented crude three-speed gears, the first of them patented by Sturmey in 1901. Bowden brought the two inventors together to form the Three-Speed Syndicate Ltd. (later Sturmey-Archer Gears Ltd.), which from 1902 to 1906 filed a series of patents relating to the first practical bicycle hub gears, which were a huge commercial success. In 1912 they invented the first motorcycle hub gear.

Collapsible scooter for the disabled

On July 5, 2001, John Scott (England) filed a patent for a compact four-wheel scooter with a seat and handlebars that fold into the vehicle body for transportation, enabling disabled people to take the scooter with them on holidays or long-distance visits.

Railway semaphore signals

Early railway signals were simply signals made by hand using flags or lamps. Where permanent signals were required, flags, balls, disks or lamps, erected on posts, would be raised, and from these evolved the semaphore signal. The first semaphores, proposed by John Rastrick, were installed by Sir Charles Hutton Gregory (both England) at New Cross Gate in London, England, in 1841.

Vehicle passenger simulator

In 2002 Alan Driver and Ken Osbourn (both England) invented a model of the human head and shoulders that could be fixed to the passenger headrest of a vehicle to give the impression that there was someone sitting in the passenger seat, thus increasing the security of lone drivers. It consisted of two sheets of material shaped like the side and front profile of a man, which slotted together at right angles to make a quasi-3D model of a person.

Opposite top Interior of the Northern Pacific Railway's luxury train *North Coast Limited* (April 1900) **Above** Pierre Vivant's traffic light sculpture at Westferry Circus, Canary Wharf, London (1999)

Traffic lights

The first electric traffic lights, comprising red and green lights and a warning buzzer, were invented by Garrett Augustus Morgan (U.S.) and installed at the corner of Euclid Avenue and 105th Street in Cleveland on August 5, 1914. The first three-color traffic lights (manually operated) were installed in New York City in 1918, but neither of these inventions was patented. Morgan filed the first patent for traffic signals in February 1922, for a system that used semaphore arms as well as red and green lights. In 1963, shortly before his death, Morgan was granted an award by the American government for his contribution to road safety.

Chapter Four

IN THE OFFICE

Air conditioning is now an essential element of many homes and cars worldwide, but for many years after its invention "air-con" was used purely as an industrial tool for the regulation of temperature, dust, and humidity in factories.

AIR CONDITIONING

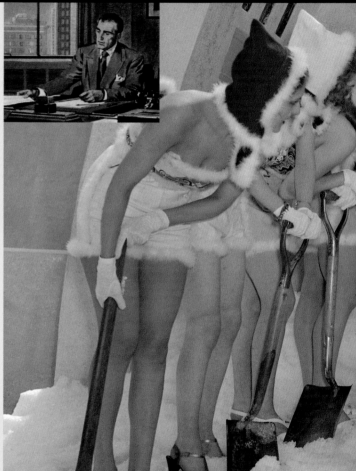

Top and Above left Advertisement for Carrier air conditioning

Artificial ventilation and air-cooling systems date back to the ancients, who hung wet mats over doorways and designed buildings so that air entered after passing through cooling fountains outside. By the 19th century fans were being used to cool air by passing it over ice, but the first scientifically designed air conditioning system was not invented until 1902, by Willis H. Carrier (U.S.).

Carrier graduated from Cornell University in 1901 with a degree in electrical engineering, and in July that year began working for the Buffalo Forge Co. in Buffalo, New York. Within six months he was put in charge of what was to become the company's research and development laboratory, where one of his first projects was to ascertain how much heat the air would absorb if circulated over a system of heating coils; his findings were to save the company thousands of dollars in heating bills. Another early project involved not heating but cooling. The Sackett-Wilhelms Lithographing and Publishing Co. in Brooklyn was having problems with colors blurring during the printing process because the paper would expand and contract with changes in temperature or humidity. On July 17, 1902, Carrier completed his designs for the world's first air conditioning unit, a 30-ton machine that precisely controlled the temperature and humidity of the print room.

Carrier continued to develop his invention, and two years later, on September 16, 1904, he filed a patent (granted 1906) for: "Apparatus for treating air"—the first water-spray air conditioning system. Air was drawn into the unit by a fan, cooled (or warmed) by the spray, and then passed through a series of baffles to remove the water and any impurities; the water was then recirculated while the conditioned air passed out of the unit to control the climate of the factory. It was not until the late 1920s that the Carrier Engineering Corp. and other companies began manufacturing air conditioning units for domestic residential use.

The inventor: Willis H. Carrier

1876 Born Willis Haviland Carrier on November 26 near Angola, New York, the son of a postmaster turned farmer

1890–94 Attends Angola Academy

1894–96 Works as a teacher

1896 Attends Central High School in Buffalo

1897 Awarded two scholarships and enrolls at Cornell University

1901 Graduates from Cornell with a degree in electrical engineering. Begins working for Buffalo Forge Company. Presents a paper that so impresses the company's executives they make him head of what becomes a research and development laboratory

Did you know?

The term "air conditioning" was coined in 1906 by textile engineer Stuart W. Cramer (U.S.), who improved Willis H. Carrier's machine for use in cotton mills by the addition of a dust filter.

Legionnaire's Disease is so called because it was first identified after an outbreak at an American Legion convention in Philadelphia in July 1976. Legionnaire's is a sometimes fatal disease affecting the lungs and is caused by bacteria that can

breed in water supplies and are sometimes spread by air conditioning systems.

Above Willis H. Carrier holds a thermometer inside an igloo display that was created to demonstrate air conditioning at the St. Louis World's Fair. The interior of the temperature-controlled igloo was kept at 68° Fahrenheit (April 25, 1939)

1902 Invents the world's first scientific air conditioning system on behalf of Buffalo Forge Company for Sackett-Wilhelms Company

1904 Files patent on September 16 for the first water-spray air conditioning system (granted January 2, 1906)

1908 Buffalo Forge Co. establishes Carrier Air Conditioning Co. a wholly owned subsidiary

1911 Publishes the *Carrier Psychometric Chart,* an air humidity graph that enables precise calculations to be made for air conditioning installations

1915 Carrier, J. Irvine Lyle (U.S.), and associates establish the independent Carrier Engineering Corporation with Carrier as president

1927 The company develops home air conditioning units

1928 Establishes the Carrier-Lyle Corporation as a subsidiary of Carrier Engineering to cater to the residential market

1930 Two other companies merge with Carrier Engineering Corp. to form Carrier Corp., with Carrier as chairman of the board

1939 The corporation develops an air conditioning system for skyscrapers

1950 Dies of a heart condition on October 7, in New York City, aged 73, having obtained more than 80 patents relating to air conditioning

1965 Four huge Carrier air conditioning units are installed in the Houston Astrodome

The Anglepoise lamp was invented by automotive engineer George Carwardine and became a design classic of the 20th century. With the launch of the Type 3 in September 2003, Carwardine's classic was updated for the 21st century.

The inventor: George Carwardine

1887 Born in Bath, England, the second youngest of 12 surviving children

1901 Leaves Bath Bluecoat School, aged 14, intending to become a minister in China as a missionary (ill health prevents him doing so)

1901–05 Apprenticed to Whiting Auto Works in Bath

1906–12 Works in various engineering workshops while studying for further academic qualifications

1912 Becomes a charge hand

ANGLEPOISE LAMP

George Carwardine (England) intended to become a church missionary, but ill health prevented him from doing so. Instead he became an apprentice automotive engineer, quickly rising through the ranks to chief designer before setting up his own company, Cardine Accessories, c. 1924. His area of expertise was car suspension systems, which gave him a highly specialized knowledge of springs, and it was this knowledge that led to the development of the Anglepoise lamp.

During the 1920s Carwardine began to develop the concept of an apparatus that could move in three planes, but that would be so well balanced that it would remain static in any chosen position. At that stage he had not considered an application for the idea and was too busy to develop it any further. Then, in 1931, he became a freelance consulting engineer, which gave him more time to devote to his equipoising concept. He realized it could be used to make an extremely versatile lamp and designed a system in which four springs were equipoised to provide the required balance. He patented this invention in 1932 as: "Improvements in Elastic Equipoising Mechanisms" and in 1933 he began to manufacture the model 1208 lamp, using springs made-to-order by specialist spring manufacturers Herbert Terry & Sons (England). In February 1934 Carwardine signed a licensing agreement with Terry's for the commercial production of the sprung lamps, which at that stage he called "equipoise lamps." He then worked with Terry's to develop an improved three-spring lamp, which he patented later that year.

When Carwardine realized that his equipoise lamp would be just as useful in the home or the office as in the factory he designed a smaller, lighter version known as the 1227. It is now considered to be the classic Anglepoise lamp. It was born in 1934 but not officially named until 1947—Carwardine was unable to use "equipoise" as a trademark because it was an existing word, so he registered the now-familiar name of Anglepoise.

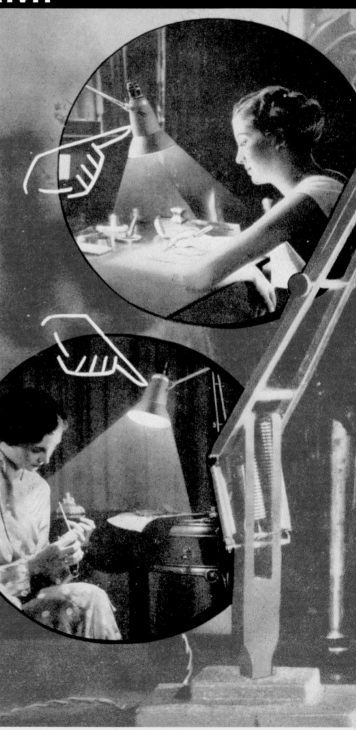

at Horstman Car Company, where he later becomes chief designer and works manager

c. 1924 Sets up a business, Cardine Accessories, designing and patenting a number of automotive innovations including rebound governors, shock absorbers, roll cheeks, and suspension moderators

1931 Becomes a freelance consulting engineer and inventor

1932 Files a patent on July 4 for the first Anglepoise lamp

1933 Produces the first Anglepoise lamp, known as model 1208

1934 Sells the manufacturing rights to Herbert Terry & Sons, England. Patents a "three-spring" Anglepoise, which leads to the classic model 1227

1947 Registers the trademark Anglepoise (U.S. reg. 1950)

1948 Dies

1951 After acquiring manufacturing rights for continental Europe and the U.S., Jac Jacobsen (Norway) successfully launches the Anglepoise in the U.S. under the name Luxo

2003 The Anglepoise Type 3 lamp is launched

Below "The need was for a light that was *instantly* adjustable....Terry's have solved this need by introducing *the Anglepoise*...!" Advertisement for Terry's Anglepoise Lamps (c. 1934)

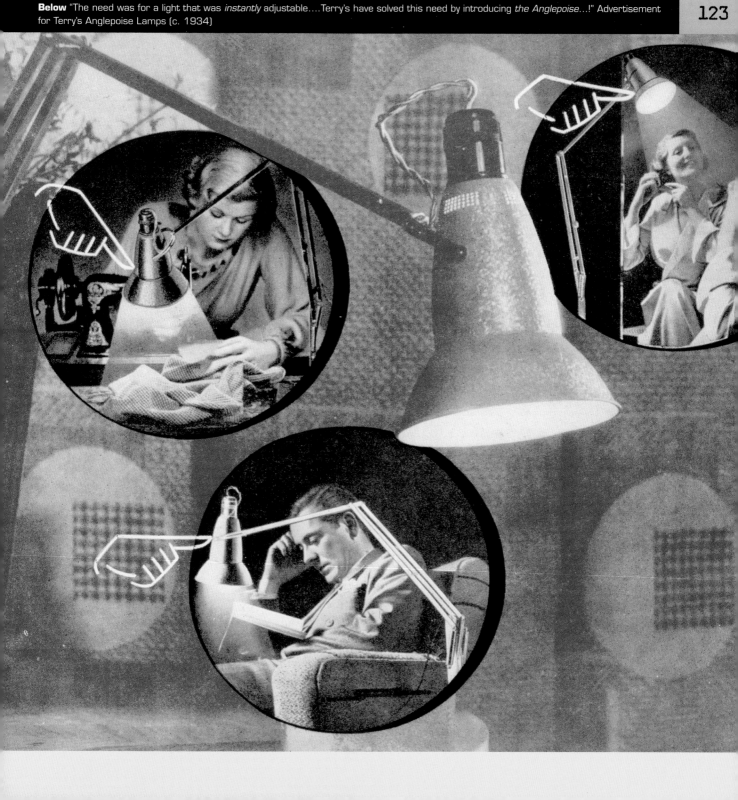

The first practical typewriter was invented by Pellegrino Turri in Italy in 1808, but the QWERTY keyboard, an essential characteristic of 20th-century typewriters, was not invented until 63 years later, by Christopher Latham Sholes.

TYPEWRITER

As with many other inventions, there are so many precursors to the modern typewriter that it is difficult to name a single inventor. The first patent for a typewriting machine was granted by Queen Anne to waterworks engineer Henry Mill (England) in 1714, for: "A machine capable of replacing handwriting by the printing of letters similar to those used in print shops." But Mill never built his machine, and it was 1808 before the first practical typewriter was invented, by Pellegrino Turri (Italy), whose first name is recorded with various spellings. Turri invented his machine as a writing aid for his friend Countess Carolina Fantoni, who was blind; 16 of her typed letters and a typed essay, dating from 1808 to 1810, are preserved in the town of Reggio, Italy.

It seems likely that Turri's machine employed a plunger for each letter, which would have been pressed directly onto the paper, and that the first typewriter to use rods converging on a single point was invented by Xavier Progin (France), and patented in 1833. The first mass-produced typewriter was invented by a pastor, Malling Hansen (Denmark), in 1865, first produced in 1870, and marketed as the *skrivekugle*, or "writing ball." Hansen's *skrivekugle* was hugely successful, and was sold across the world, but it was destined not to become the standard blueprint for typewriters.

In 1867 journalist and newspaper editor Christopher Latham Sholes, with Carlos Glidden and Samuel Soule (all U.S.), developed an idea by John Pratt (England) and filed a patent for a "type writing machine." Sholes patented improvements in 1871 and went on to invent the first typewriter to use the now ubiquitous QWERTY keyboard —the precursor of all modern typewriters and computer keyboards. In March 1873 Sholes and his business partner James Densmore (U.S.) completed a deal with the Remington Small Arms Co. (U.S.) to mass-produce the machine as the Sholes & Glidden Type-Writer. The name was changed in 1876, and Sholes's machine went on to achieve fame as the Remington No.1.

See also: Carbon paper, page 148

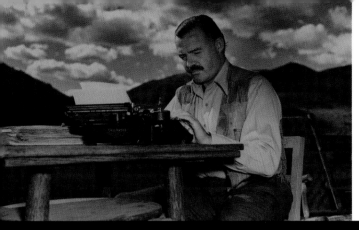

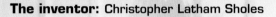

Above American writer Ernest Hemingway (1889–1961) works at his typewriter while sitting outdoors in Idaho. (Hemingway disapproved of this photo, saying, "I don't work like this") (October 1939) **Below** A 1950s shopper is overwhelmed by the choice of typewriters

The inventor: Christopher Latham Sholes

1819 Born on February 14 in Pennsylvania, the son of Orrin Sholes, a farmer. (His father received the farm as a reward for serving in the War of 1812.)

1833–37 Serves a journalistic apprenticeship with the *Danville Intelligencer*

c. 1838 Moves to Wisconsin, where he works as a printer for the *Wisconsin Democrat*

1839 Becomes editor of the *Wisconsin Enquirer*

1840 Becomes editor of the *Southport Telegraph*

1844 Appointed postmaster of Southport, Wisconsin, by President James Polk

c. 1859 Designs a machine for printing the names of subscribers in the margin of newspapers (manufactured 1860)

1860 Moves to Milwaukee to become editor of the *Milwaukee News* and later the *Milwaukee Sentinel*

1863 Appointed Collector of the Port of Milwaukee by President Abraham Lincoln

1864 & 1866 With his friend Samuel Soule (U.S.), Sholes patents inventions for numbering printed pages

1867 Carlos Glidden (U.S.) shows Sholes an issue of *Scientific American* describing John Pratt's British patent relating to typewriting. Together they and Soule develop and patent a "type writing machine" (granted 1868)

1868–73 Investor James Densmore (U.S.) upsets Glidden and Soule, who leave the partnership. Sholes patents improvements including the QWERTY keyboard

1873 Densmore and Sholes strike a manufacturing deal with the Remington Small Arms Co. Remington offers cash or royalties; Sholes takes the cash, but Densmore opts for royalties and grows rich from the invention

1890 Sholes dies on February 17, aged 71, after prolonged ill health

For more than 300 years Blaise Pascal was credited with inventing the mechanical calculator. But in 1957 a German historian proved that Wilhelm Schickard had beaten Pascal to it by 18 years, with his invention of a so-called "calculator clock" in 1624.

CALCULATOR

In the 17th century John Napier (Scotland) had paved the way for the invention of mechanical calculating machines with his proposal that multiplication and division could be calculated as a series of additions and subtractions. Then, for more than three centuries, historians believed that Blaise Pascal (France) had invented the calculator in 1642, but this was disproved in the mid-20th century when historian Franz Hammer (Germany) discovered papers proving that in 1624, fully 18 years before Pascal, Wilhelm Schickard (Germany) had invented something called a "calculator clock." Schickard's machine not only predated Pascal's, it was also far more sophisticated and was capable of performing all four arithmetic operations of addition, subtraction, multiplication, and division. Having discovered Schickard's papers in 1935, Hammer published them in 1957 and Schickard was finally acknowledged as the inventor of the mechanical calculator.

The first commercially successful calculator was invented in 1820 by Charles Xavier Thomas de Colmar (France), but the evolution of calculators as we know them came when Jack Kilby (U.S.), inventor of the microchip, began developing miniature electronic calculators as a means of exploiting the microchip. In 1967 Kilby, together with Jerry Merryman and James van Tassel (both U.S.) of Texas Instruments (TI), produced the first hand-held electronic calculator and three years later TI and Canon Inc. (Japan) launched the first commercial electronic pocket calculator, the Pocketronic.

Some historians cite the Sinclair Executive, invented in 1972 by Clive Sinclair (England), as the first pocket calculator, but that depends on the size of one's pocket. The Sinclair Executive was 9.5mm thick and 140mm long and used LEDs (light-emitting diodes) to display its results, while the Pocketronic, which had a thermal paper print-out for results, was large enough to stretch the definition as well as the fabric of a pocket.

See also: Microchip, page 246; Logarithms, page 260

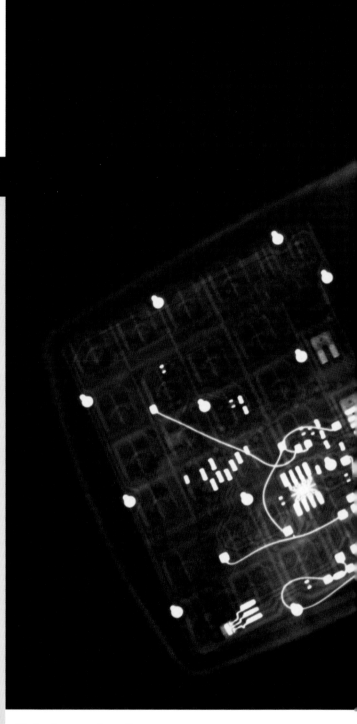

Calculator time line

c.3000 BC The abacus is invented as the first form of mechanical calculating aid. In the 21st century it is still widely used in India, China (where it is known as the *suan pan*), Japan (*soroban*) and Russian Federation (*tschoty*)

1614 John Napier (Scotland) publishes his invention of logarithms

1624 Wilhelm Schickard (Germany) invents the first calculating machine

1642 Blaise Pascal (France) invents the "pascaline," which for more than 300 years is thought to be the first adding machine

1671 Gottfried Leibniz (Germany) invents a "stepped reckoner," the first calculator

Left X-ray of calculator **Above** French mathematician Charles Xavier Thomas de Colmar, inventor in 1820 of the first commercially successful calculator (c. 1850)

Did you know?

Anthony Patera (U.S.), Associate Professor of Mechanical Engineering at the Massachusetts Institute of Technology (U.S.), declared in 1990 that: "The calculator is as necessary in class as pen and paper." His colleague David Wilson (U.S.) put the importance of calculators in even more simple terms, saying: "Calculators revolutionized calculating."

capable of extracting square roots

1820 Charles Xavier Thomas de Colmar (France) produces the first commercially successful calculator, the Thomas de Colmar Arithmometer

1885 Dorr Eugene Felt (U.S.) invents the "comptometer," the first key-driven adding

machine, patented late the following year

1930 Vannevar Bush (U.S.) invents the first electro-mechanical calculator

1964 Sharp Corporation (Japan) invents the first all-transistor desktop calculator

1967 Jack Kilby, Jerry Merryman and James van

Tassel (all U.S.), working for Texas Instruments, produce the first hand-held electronic calculator

1970 Texas Instruments (U.S.) in conjunction with Canon Inc. (Japan) launches "Pocketronic," the first electronic "pocket" calculator

1972 Hewlett-Packard (U.S.) invents the first "scientific"

hand-held electronic calculator (i.e. capable of performing trigonometric and exponential functions as well as arithmetic functions). Sinclair Radionics Ltd (England) produces the Sinclair Executive, the first truly pocket-sized electronic calculator

1974 Hewlett-Packard invents the first programmable hand-held calculator

To name the inventor of the computer is problematic because it depends on the definition of the word "computer." However, it is much easier to define the origin of the mouse, which was invented in 1964 by Douglas Engelbart.

COMPUTER & MOUSE

Computer

A number of people have been hailed as "the inventor of the computer" since Charles Babbage (England) designed his Analytical Engine in 1835, which would have been the first programmable computer except that it was never built. Other candidates include Konrad Zuse (Germany); John Atanasoff and Clifford Berry (U.S.); Max Newman, Alan Turing, and Thomas Flowers (Britain); and John Mauchly and Presper Eckert (U.S.).

The machines of all these inventors heralded technological advances, but two things are crucial to the modern definition of a computer: It must be programmable, and it must have stored memory (RAM, or random access memory). By that definition, the computer was finally "invented" by Professor Freddie Williams and Tom Kilburn (both England) at Manchester University in 1948 when they produced "Baby," the world's first stored-program computer, which first ran on June 21, 1948, successfully performing a 17-instruction program that signaled the birth of the first true computer.

Mouse

Douglas Engelbart (U.S.) has over 40 patents to his name relating to computing, the most memorable of which, though not necessarily the most important, is the computer mouse. Engelbart wanted to find the best way to do things onscreen that a keyboard could not do. He looked into the existing ideas, which included joysticks and light pens, and combined the best features of all of them. Then, in 1964, he invented what he later patented as: "An X-Y position indicator for a display system" (filed 1967, granted 1970). This device was designed to be moved across a flat surface and contained two discs at right angles to each other, one to calculate the left-right position (X axis) and one to calculate the up-down position (Y axis) on the screen; with its compact, curved casing and tail-like wire, Engelbart's invention was soon nicknamed the mouse.

Above Kristopher Couch showing off the graphic capabilities of the

The inventors: Freddie Williams, Tom Kilburn, & Douglas Engelbart

1911 Williams is born Frederick Calland Williams on June 26 in Stockport, Cheshire, England. Studies engineering at Manchester and Oxford Universities, England

1921 Tom Kilburn is born on August 11 in Dewsbury, England. Studies at Cambridge University, England

1925 Douglas Engelbart is born on January 30 near Portland, Oregon

1942–45 Engelbart serves in the U.S. Navy as a radar technician

1946 Williams becomes professor of electrical engineering at Manchester University

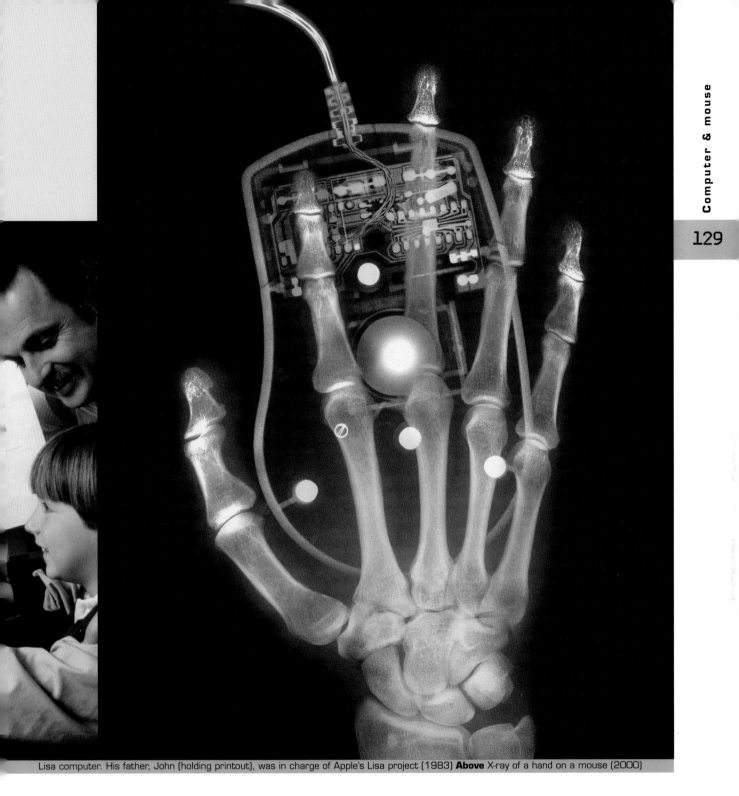

Lisa computer. His father, John (holding printout), was in charge of Apple's Lisa project (1983) **Above** X-ray of a hand on a mouse (2000)

1948 Williams & Kilburn perfect "Baby," the world's first stored-program computer

1950 Williams elected a Fellow of The Royal Society

1950s Engelbart gains a Ph.D. at Berkeley and begins teaching there

1957 Engelbart moves from Berkeley to Stanford Research Institute

1959 Engelbart is given his own lab at Stanford (later known as the Augmentation Research Center)

1962 Kilburn designs the ATLAS computer, pioneering virtual memory and multiprogramming

1963 Engelbart given funding by the Pentagon's Advanced Research Projects Agency

1964 Engelbart invents the computer mouse. Kilburn becomes professor of computer science at Manchester University

1965 Kilburn elected a Fellow of The Royal Society

1967 Engelbart files a patent for the computer mouse (granted 1970)

1976 Williams is knighted

1977 Williams dies on August 11 in Manchester, aged 66

1989 Engelbart founds The Bootstrap Institute, at Stanford University, which promotes the development of collective IQ through worldwide computer networks

2001 Kilburn dies on January 17 in Manchester, aged 79

Laszlo Bíró is so famous as the inventor of the first practical ballpoint pen that his name is often used as a generic term for all makes of ballpoint—and it is often overlooked that he was also a sculptor, painter, hypnotist, and journalist.

BALLPOINT PEN

The *idea* of the ballpoint pen was first patented in 1888 by John J. Loud (U.S.), but he could not devise a suitable ink feed, and a practical version of his pen was never made. Other patents followed, but it was not until 1945 that Laszlo Josef Bíró (Hungary–Argentina) finally managed to produce the first practical ballpoint pen; while the idea itself was not novel, he had invented a means of making it work.

Bíró was born Laszlo (sometimes recorded as Ladislao) Josef Bíró on September 29, 1899, in Hungary. He worked in Budapest as a journalist and artist, and during the 1930s became the editor of a cultural magazine. It was on a visit to the printing press that he began to wonder whether printers' ink, which was quick-drying and therefore less likely to smudge, could be used in a pen. Printers' ink was too thick to be used in a fountain pen, so Bíró conceived the idea of replacing the traditional nib with a metal ball. He patented this idea in 1938, but this early pen was impractical. At the beginning of the Second World War, Laszlo and his brother Georg fled from Hungary, first to Paris and then, in 1940, to Argentina. They continued to work on the pen, and in 1943 Laszlo filed a patent for the world's first commercially practical ballpoint pen (granted 1944).

That summer Bíró met British government official Henry Martin (England) in Buenos Aires. Martin thought that, because they would not leak at high altitudes, Bíró's pens would be useful for RAF pilots. He bought the British rights to Bíró's invention, established the Miles-Martin Pen Company, and began manufacturing the first ballpoint pens in 1944. Bíró was about to launch his pen in America through Eversharp, but he had inadequate patent protection and was beaten to it by businessman Milton Reynolds (U.S.), who launched America's first ballpoint, the Reynolds Rocket, in 1945.

Did you know?

Bíró met President Justo of Argentina while both men were on vacation in Yugoslavia. Justo, who was intrigued by Bíró's invention, invited him to live and work in Argentina, an invitation that Bíró accepted after the outbreak of the Second World War.

Above left Laszlo Bíró writing with one of his pens **Above right** Colored electron scanning micrograph of the ball at the nib of a ballpoint pen

Ballpoint pen time line

1888 John J. Loud (U.S.) is granted the first patent for a ballpoint pen, but cannot devise a suitable ink feed

1899 Laszlo Bíró is born on September 29

1938 Bíró patents a prototype ballpoint pen

1943 Bíró (now working in Argentina) patents the first practical ballpoint pen and sells the British rights to Henry Martin (England)

1944 The Miles-Martin Pen Co. manufactures Britain's first ballpoint pens, for the RAF (first sold to the public at Christmas 1945)

1945 Eterpen Co. (Argentina) produces the first commercially available ballpoint pens (under license from Bíró in Buenos Aires, Argentina).

Businessman Milton Reynolds (U.S.) employs William Huenergardt (U.S.) to copy Bíró's idea. The result is the Reynolds Rocket, America's first ballpoint pen, patented in 1945 and first sold by Gimbel's of New York the same year

1953 Baron Marcel Bich (France) acquires the rights to Bíró's patent and manufactures the first disposable ballpoint pen, marketed as the BiC Crystal (introduced to Britain in 1958)

1978 Gillette (U.S.) invents the first erasable ballpoint pen, known as the Eraser Mate

1985 Bíró dies on October 24, aged 86, having also invented a heatproof tile and a high-security lock

Bette Nesmith, later Nesmith Graham, had an enormous influence on two very different areas of life: office work and pop music. In 1942 she gave birth to Mike Nesmith, later of the pop group The Monkees, and c. 1951 she invented Liquid Paper.

CORRECTION FLUID

Like many inventors, Bette Nesmith Graham did not grow up expecting to be an inventor—she had her sights set on becoming an artist. Born Bette McMurry on March 23, 1923 in Dallas, Texas, she dropped out of high school at the age of 17 and managed to get a job as a secretary at a Dallas law firm even though she couldn't type (the firm later sent her to secretarial school). She married Warren Nesmith in 1940 and two years later, on December 20, 1942, she gave birth to Mike, who would later find fame as the guitarist for the pop group The Monkees.

The Nesmiths divorced after the Second World War and Bette, now a single parent, began working as secretary to W.W. Overton at the Texas Bank and Trust, where she had the brainwave that was to make her fortune. She noticed that the bank's sign writers would simply paint over any mistakes, and she wondered if she could cover up typing errors in the same way. As an artist she was used to handling paint, and she mixed up a small bottle of water-based paint to match the color of the company stationery; she then began using a watercolor brush to paint over any typing errors. Overton didn't object, and soon other secretaries began to use her invention, which she called Mistake Out.

Nesmith improved the formula to make it faster drying, and began mixing ever larger quantities using the electric mixer in her kitchen; Mike and his friends would bottle it in the garage. In 1956 she offered to sell the rights for her invention to IBM, who rejected it. Undeterred, she set up her own company, Mistake Out Co., and began selling correction fluid herself. After being fired from Texas Bank and Trust, she and her second husband, Bob Graham, devoted more time to the business (now called Liquid Paper), and it began to grow exponentially. Nesmith used part of her resulting fortune to set up two foundations for women, and when she died in 1980 she willed half her estate to charity.

Left Correcting mistakes on a typewriter is painstaking work **Right** A bottle of correction fluid, also known as Liquid Paper or White-Out

Correction fluid time line

c. 1951 Bette Nesmith invents a correction fluid that she calls Mistake Out

1956 Nesmith offers her invention to IBM, which rejects it. She sets up her own company, Mistake Out Co.

1958 Sales reach 100 bottles per month

1962 Bette Nesmith marries Bob Graham

1968 Mistake Out Co. is renamed Liquid Paper and moves to a 1,200 sq yd (1,000 sq m) factory with fully automated production

1975 Liquid Paper moves into a 3,900 sq yd (3,250 sq m) international headquarters in Dallas, Texas. Production reaches 25 million bottles per year, exported to 31 countries

1979 Graham sells Liquid Paper to Gillette for $47.5 million and royalties on every bottle until the year 2000

1980 Graham dies at the age of 56, six months after selling her business. She wills half her estate to her son Mike and half to charity

Did you know?

At the original auditions for The Monkees, Stephen Stills (who later came to fame with Crosby, Stills & Nash) was rejected because he had bad teeth. Mike Nesmith has since been described as the group's "only real musician." Stills didn't get the part in The Monkees, but his friend Peter Tork did after Stills recommended him to Bob Rafelson and Bert Schneider (the producers). However, Stills did get a part behind the scenes playing guitar on their album *The Birds, The Bees, And The Monkees* (1968).

The accidental discovery of a glue that wouldn't stick, coupled with a conscientious chorister who didn't want to damage his hymn book, led to the invention of the Post-it Note, which has since become an indispensable part of office life.

POST-IT NOTE

The Post-it Note was invented in 1974 by Arthur Fry (U.S.), an employee of the 3M Corporation. Necessity is said to be the mother of invention, but in fact the Post-it Note fulfilled a need that most people weren't aware they had—and its invention developed from an earlier serendipitous discovery by a colleague of Fry's, research chemist Spencer Silver (U.S.). Silver had been researching adhesives, but instead of coming up with a strong glue, he came up with the exact opposite. Silver's patent, filed on behalf of 3M in 1970, states that: "This invention relates to inherently tacky, elastomeric, solvent-dispersible, solvent-insoluble, acrylate copolymer and a process of preparing the copolymer"—in other words, a glue that didn't stick.

Significantly, this weak glue was reusable, and it did not leave a sticky residue when it was removed—paradoxically, its weakness was its strength, because it was so weak that it did not even damage paper fibers after being stuck to a page. Silver thought that there must be a practical use for his invention, and even proposed a noticeboard covered with the glue to which notes could be stuck. But it wasn't until 1974 that a use was found: Art Fry, a member of a church choir, used Silver's "unglue" to stop the page markers from falling out of his hymn book by coating the top edge of each marker with the glue. Fry said later: "I needed a bookmark that would stay put, yet could easily be removed without damaging my hymnal."

For three years, Fry developed his idea to the point that it was marketable. At one point, he wrote a note to a colleague on one of his bookmarks, and it came back with the response written below it on the same marker, prompting 3M to test-market Post-it Notes in 1977 as temporary notes rather than as bookmarks. Officially launched in 1980, Post-it Notes are now a worldwide phenomenon, to the delight of Art Fry: "It is like having your children grow up and turn out to be happy and successful."

The inventors: Spencer Silver & Arthur Fry

1931 Arthur Fry is born on August 19 in Owatonna, Minnesota

1941 Spencer Silver is born on February 6 in San Antonio, Texas

1953 Fry begins working part time in the industrial tape division of 3M while studying at the University of Minnesota

1955 Fry graduates with a degree in chemical engineering and takes full-time employment with 3M, where he works until 1992

1962 Silver graduates from Arizona State University with a B.S.

1966 Silver graduates from the University of Colorado with a Ph.D. and begins working for 3M

1969 Silver invents the weak glue that leads to the invention of Post-it Notes and files a patent in 1970 for "Acrylate copolymer microspheres"

1974 Fry invents the Post-it Note

1980 Post-It Notes launched in the U.S. and Canada

1981 Post-It Notes named 3M's Outstanding New Product

1992 Fry retires and becomes a consultant to 3M

1995 3M wins America's National Medal of Technology, with Post-it Notes mentioned in the citation

1996 Silver retires from 3M with 23 U.S. patents to his name

Did you know?

3M was originally formed in 1902 as the Minnesota Mining and Manufacturing Co., to quarry corundum for use in abrasive grinding wheels. In 1905 the company began making sandpaper, and in 1916 opened a quality control laboratory, which eventually came to be used for research and development. As a result the company began to diversify, and signaled the move away from its mining origins by adopting the abbreviation 3M.

Did you know?

One of the earliest known references to the use of bookmarks is Albrecht Dürer's 1526 painting *Portrait of Erasmus,* which shows the eponymous Erasmus using a scrap of paper as a bookmark.

The engineering and production departments at 3M told Arthur Fry that Post-it Notes would "pose considerable processing measurement and coating difficulties," but Fry was undeterred. His response was: "Really, that is great news! If it were easy, then anyone could do it. If it really is as tough as you say, then 3M is the company that can do it."

Opposite Art Fry demonstrates the "Eureka!" moment that led to the birth of Post-It Notes **Below** Post-It Notes are now ubiquitous in offices, but there can be too much of a good thing

At once a design classic and a desktop accessory that has remained popular even in the computer age, the Rolodex brand card file was invented by Arnold Neustadter in 1952 and successfully marketed for the first time in 1958.

ROLODEX

Arnold Neustadter (U.S.) was a man with a mission—he wanted to create the perfect "dex": he used "dex" as a suffix for numerous inventions, all of them relating to office desks and to the storage and organization of information. During the 1930s he invented his first "dex," a telephone directory that he called the Autodex. The user moved a telephone-type dial to the appropriate letter of the alphabet, and the spring-powered Autodex would pop open at the relevant page. Compared with what was to come, the Autodex was a modest success, but it was enough to encourage Neustadter to establish a company, which he called Zephyr American, in order to manufacture and market his inventions.

Autodex was followed by a spillproof inkwell called the Swivodex, and then came the Punchodex, for punching holes in paper, and the Clipodex, a device that clipped to the knee of a typist or secretary to help in taking dictation. Then, in the 1940s Neustadter invented a revolving card cylinder that he called Wheeldex. Office life changed forever in 1952 when he and engineer Hildaur Nielson (U.S.) improved the Wheeldex, which evolved into the classic Rolodex: a card filing system that rotated on a cylinder, giving instant access to any one of scores of slotted cards that snapped into place on the cylinder hub.

The simple, elegant design of the Rolodex was an engineering and aesthetic mini-masterpiece. The card-wheel was suspended over the base on a tubular steel, cantilevered frame, and the Rolodex was ergonomically designed with not one but two perfectly sized knurled knobs to rotate the cylinder, making the Rolodex as easy to use for the left-handed as for the right. Despite its eminent practicability and up-to-the-minute fifties design, sales were slow at first. Then a combination of good marketing and its use by Jack Lemmon in the film *The Apartment* made the Rolodex de rigueur for office desks, and it has remained popular ever since.

Above A secretary flips through a Ferris wheel office file on display at the 41st National Business Show, New York (October 1949)
Opposite A well-thumbed Rolodex can be overwhelmingly handy

Did you know?
Not only has Arnold Neustadter's 1950s invention survived into the computer age, it has actually become part of it—there is a computer program that prints out information for Rolodex card files.

The inventor: Arnold Neustadter

1910 Born in Brooklyn, New York. He later attends New York University and then works at his father's box factory

1938 Establishes Zephyr American (later Rolodex) to manufacture and market his inventions

1930s–40s Invents Autodex, Swivodex, Punchodex, Clipodex and Wheeldex

1952 Improves Wheeldex to become Rolodex

1958 Successfully markets Rolodex

1970s Sells the company to Insilco (International Silver Co., USA) and retires, devoting his remaining years to philanthropy and to his collection of antique paperweights

1996 Dies on 17 April

1997 The Rolodex division of Insilco is bought by Newell Rubbermaid (USA)

Like the safety pin, the paper clip is a gloriously simple invention made from a single piece of wire, and like the safety pin it was invented in several places at almost the same time, although the credit usually goes to Johann Vaaler.

PAPER CLIP

The paper clip is a relatively humble invention, but billions of them are used every year in offices around the globe. As with so many inventions, it was the result of an evolutionary process, and therefore it is difficult to name a single "inventor." In 1896 Matthew Schooley (U.S.) filed a U.S. patent for a bent wire "paper clip or holder" that looked very much like a modern paper clip, but even this early patent stated: "I am aware that prior to my invention paper-clips have been made somewhat similar to mine in their general idea..."

Johann Vaaler (Norway) is generally accepted as the inventor of the paper clip, although he filed his patent three years after Schooley, in 1899 (granted 1900). One reason that Vaaler is usually named as the inventor may be that Vaaler's clip was flat (as is the standard modern clip), while Schooley's, although a very similar shape, was raised in profile like the first coil of a spring in order to bind the papers without "puckering or bending" them—this, of course, would depend on the thickness of the sheaf of papers and proved to be an unnecessary embellishment.

However, Vaaler's design was no more the prototype of the modern paper clip than Schooley's: The defining characteristic of the standard modern clip is the loop within a loop, but Vaaler's had only a loop and a "leg." The double oval paper clip, now the standard design, is thought to have its origins with Gem Manufacturing Ltd. (England), although the first documentary evidence of this design comes in a U.S. patent for a paper clip-making machine, filed in 1899 by W.D. Middlebrook (U.S.). The double oval design was never explicitly patented (Middlebrook patented only the machine), but it is shown in the patent drawing as the end product of Middlebrook's machine, and was therefore clearly already in use by 1899, weakening Vaaler's claim. But whatever its provenance, and whoever the inventor, "Gem" is now the name used to describe the classic loop-within-a-loop paper clip.

Paper clip time line

1896 Matthew Schooley (U.S.) files a patent for a "paper clip or holder" (granted 1898)

1899 Johann Vaaler (Norway) files a patent for a paper clip; because Norway has no patent laws he files his original application in Germany (granted 1900). W.D. Middlebrook (U.S.) patents a "machine for making paper clips." The patent drawings clearly show a double loop "Gem" paper clip as the end product of the machine

1900 Cornelius J. Brosnan (U.S.) files a patent for a paper clip that he calls the Konaclip, a loop-and-leg clip that has since been described as "the first successful bent wire paper clip"

Left Modern sculpture adorns a steno pad **Below** Paper clip chains keep any from being separated from the group

139

1901 Vaaler is granted a U.S. patent for his paper clip

1907 Gem Manufacturing Ltd. (England) advertises a slide-on paper clip that it declares will "hold securely your letters, documents, or memoranda without perforation or mutilation until you wish to release them"

1908 The Gem is advertised as America's most popular clip, "the only satisfactory device for temporary attachment"

1921 Clarence Collette (U.S.) is granted a U.S. patent for a paper clip with "sharply pointed projections for penetrating and engaging the sheet material." It is intended to overcome the problem of paper clips slipping off the

papers they are holding, but in fact exacerbates the problem of damage to said papers

1925 Collette is granted a U.S. patent for an improved design with ridges, rather than projections, to grip the paper

1934 Henry Lankenau (U.S.) files a patent for a design that survives today, described thus in the patent: "One end portion

of the clip consisting of a single loop of rectangular form and the opposite end portion consisting of a double loop and being V-shaped in a lengthwise direction." Lankenau's clip is shrewdly marketed as the Perfect Gem, but comes to be known as the Gothic-style paper clip

1950s Colored plastic paper clips are introduced

Alexander Graham Bell achieved fame as the inventor of the telephone, but in fact what he invented was the first *commercially practical* telephone: Johann Philipp Reis, Antonio Meucci, and others had already invented working telephones before Bell.

TELEPHONE

> **Did you know?**
> At first, Alexander Graham Bell's rival, Elisha Gray, did not challenge Bell's patent, having been advised that the telephone was an insignificant distraction from Bell and Gray's original goal of improving the telegraph. After the importance of the telephone became clear, Bell suffered more than 600 lawsuits, including many from Gray, none of which was upheld.

The first public demonstration of a working telephone was made by Johann Philipp Reis (Germany) in 1860, but Reis's invention used an intermittent current, giving a spasmodic signal at the receiver, and never progressed beyond the experimental stage. In 1871 Antonio Meucci (Italy–U.S.) filed a caveat (a registration of an idea in progress pending a full patent application) for a telephone that he claimed to have invented in 1849/1850 and to have demonstrated between December 1860 and December 1861, but he never filed a full patent for his invention.

In 1873 Alexander Graham Bell (Scotland–U.S.), the son of an elocution teacher, was appointed professor of vocal physiology at Boston University, where he taught deaf students using the Visible Speech System invented by his father. His interest in acoustics led to the invention of his telephone during 1875 and 1876, using *variations* in an electric current, which Bell called "undulatory currents," as opposed to Reis's intermittent current. Bell filed his patent application on February 14, 1876, just hours before his rival, Elisha Gray (U.S.). Bell's patent was granted on March 7, and three days later he made his famous first telephone call, saying: "Mr. Watson, come here, I need you" (also quoted as "Mr. Watson, come here, I want you"). Having failed to sell the patent rights to the Western Union Telegraph Company in 1877, Bell and his financial backers formed the Bell Telephone Company, which eventually became one of the largest companies in the world.

Bell's work as a vocal physiologist teaching the deaf and speech impaired remained his first priority, and in order to pursue this (and his interest in flying) in 1880 he resigned from what was by then the American Bell Telephone Company. Bell maintained that he did not really comprehend the science behind his invention, only that it worked, and he once wrote to his wife: "I think I can be of far more use as a teacher of the deaf than I can ever be as an electrician."

Above Alexander Graham Bell and Thomas Watson with an early telephone (c. 1887) **Oppposite** Joan Crawford has telephone trouble

The patentee: Alexander Graham Bell

1847 Born March 3 in Edinburgh, Scotland, the son of an elocution teacher, and later educated at McLauren's Academy, Edinburgh

1860 Graduates from the Royal High School

1870 Bell's family immigrates to Ontario, Canada, on the advice of doctors after Bell's two brothers die of tuberculosis

1872 Moves to the U.S.

1873 Appointed professor of vocal physiology at Boston University

1875–76 Invents the first practical telephone, which he patents in 1876

1877 Forms the Bell Telephone Company

1880 Resigns from what is now the American Bell Telephone

Did you know?

Bell was supported in his work by Gardiner Greene Hubbard and Thomas Sanders (both U.S.), the fathers of two of his deaf pupils. Bell later married one of the pupils in question, Mabel Hubbard.

It is said that Bell used his metal detector (1881) to search for the assassin's bullet in President James Garfield, who was shot by Charles Guiteau on July 2, 1881. Bell could not find the bullet, but reportedly detected the metal bedsprings beneath the President, who died on September 19.

Company. Establishes the Volta Laboratory for research, invention, and work for the deaf. Invents the Photophone, a machine capable of transmitting speech using light rays

1880–81 Invents the Graphophone, an improvement on Edison's phonograph that features an engraving recording head. Among the first words recorded on the prototype (lodged at the Smithsonian Institution) are: "I

am a Graphophone and my mother was a Phonograph."

1881 Invents a primitive metal detector, which he calls an induction balance, but does not develop it. (The metal detector was reinvented on the same principle by Gerhard Fischar [Germany] in 1925.) After the death of his newborn son from respiratory failure, he designs a prototype iron lung, which he calls a "vacuum jacket" *See also: Iron lung, page 226*

1882 Becomes a U.S. citizen on November 10

1883 Helps found journal *Science*

1888 Helps found the National Geographic Society

1890 Founds the American Association to Promote the Teaching of Speech to the Deaf (now the Alexander Graham Bell Association for the Deaf)

1903 Makes the first published proposal to use radium

inserts, rather than external rays, in the treatment of cancer (published in *Science*)

1904 Patents a "compound cellular aerial vehicle"

1907 Helps found the Aerial Experiment Association to advance aviation technology

1922 Dies on August 22 in Nova Scotia, Canada, aged 75, having taken out 30 patents during his lifetime

In an age when faxes are transmitted using telephone lines, it is surprising to find that the first fax machine was invented three decades earlier than the telephone—the first faxes were actually transmitted by telegraph.

The inventors: Alexander Bain & Giovanni Caselli

1811 Alexander Bain is born near Thurso, Scotland. He later trains as a watchmaker and moves to London

1815 Giovanni Caselli is born in Siena, Italy. He later becomes an abbot

1840 With John Barwise (Britain), Bain patents the first electric clock (granted 1841)

1843 Bain files a patent containing the first description of a facsimile machine

FAX MACHINE

On November 27, 1843, Alexander Bain (Scotland) filed a patent containing the first description of the facsimile (or fax) machine, and outlined the basic principle that is still used in modern fax machines—the transmission of an electrical signal to indicate whether a given part of the document is black or white. Bain proposed using synchronized pendulums to "map" a document written in an electrically conductive material, but he never actually made a transmission. In 1848 physicist Frederick Bakewell (England) patented an improved version of Bain's machine, and made the first public demonstration of a fax transmission at Britain's Great Exhibition of 1851.

Bakewell's machine worked, but it was not a commercial success, and it was another decade before the first truly practical fax machine was invented, by Giovanni Caselli (Italy), an abbot who was granted a patent for his "pantelegraph" in 1861. The pantelegraph incorporated ideas from both Bain's and Bakewell's patents, and worked on a similar principle—but it was efficient enough for the French government to set up the first commercial fax line, from Paris to Lyons, in 1865 (the line was later extended to Marseilles).

Despite Caselli's initial success, use of the fax machine did not become widespread because of vested interests in conventional telegraphy using Morse code. After the Franco-Prussian War made his French system infeasible, Caselli's fax machine fell into relative obscurity until the 20th century, when Alexander Korn (Germany) developed the concept of photoelectric scanning. The principle remained the same—the photoelectric scanner would "map" the black and white areas of the document—but ordinary documents could now be sent, regardless of their electrical conductivity or resistance, and fax machines no longer relied on pendulums. These advantages, plus the fact that the telegraph had given way to the telephone, led to the more widespread use of fax machines, first by newspapers and then by businesses in general.

Did you know?
The word "facsimile" is a conflation of two Latin words, *fac simile,* meaning "to make the same."

1844 Bain patents navigational instruments for ships

1861 Caselli is granted a British patent for his pantelegraph

1865 The first commercial fax service is set up in France using Caselli's invention

1873 Bain is granted an annual pension of £80 after fellow inventors petition British prime minister William Gladstone

1877 Bain dies in Kirkintilloch, Scotland, having patented 10 inventions (mostly relating to telegraphs and timepieces)

1891 Caselli dies in Florence, Italy

Above Bain's facsimile machine (photographed in 1850)
Top right Cover of *The New Yorker* (December 1992)

The first photocopier, invented in 1903, in fact took photographs of the documents to be copied, but it was a long, slow process and did not catch on. The modern, electrostatic photocopier was invented by Chester F. Carlson in 1938.

The inventor: Chester F. Carlson

1906 Born on February 8 in Seattle, Washington, the son of a barber

1930–33 Graduates in physics from the California Institute of Technology. Works as a research engineer for Bell Telephone Laboratories before

losing his job as a result of the Depression

1934 Begins working in the patent department of electronics firm P.R. Mallory & Co., and enrolls for evening classes at the New York Law School

By the time Chester F. Carlson (U.S.) was born in 1906, various methods for copying documents had been invented, including carbon paper and the blueprint. In 1903 George C. Beidler (U.S.) invented the first photocopier, known as the Rectigraph, but this process involved developing photographic prints of documents and was not a success. It was not until Carlson invented what he called "electron photography" in 1938 that documents could be copied electrostatically.

Carlson worked as a research engineer for Bell Telephone Laboratories until he lost his job as a result of the Depression. He then took a job first for a patent lawyer and then in the patent department of New York electronics firm P.R. Mallory & Co., where he turned his mind to inventing a copier to speed up his patent work. Researching at the New York Public Library, he discovered the principle outlined by Paul Selenyi (Hungary) that the electrical conductivity of certain materials is affected by light. He applied Selenyi's principle to copying, and experimented with the use of altered conductivity to create a fixable "shadow" of whatever was to be copied. On September 8, 1938, he filed a patent for "electron photography," and the following month he made his first successful use of the process, copying the date and place of the experiment from a glass plate onto waxed paper: "10-22-38 Astoria."

Carlson approached more than 20 companies with his invention, but none of them was interested until, in 1944, the Battelle Memorial Institute, of Columbus, Ohio, agreed to develop the idea under a royalty agreement. The process was developed by Roland M. Schaffert (U.S.), and the manufacturing rights were eventually sold to the Haloid Corporation (U.S.), which in 1948 renamed Carlson's invention "xerography" (from the Greek *xeros*/dry and *graphein*/to write). Haloid produced the first successful xerox machine in 1959, and its copiers proved so successful that the company later changed its name to the Xerox Corporation.

1938 Makes his first copy using "electron photography" and files a patent relating to the process

1939 Files a patent relating to an improved process that he now calls "electrophotography" (granted 1942). Graduates with a law degree from New York Law School

1944 Receives a patent relating to improved apparatus for electrophotography. The Battelle Memorial Institute of Columbus, Ohio, agrees to develop the idea

1947 Battelle persuades Haloid Corp. (U.S.) to sponsor further research in return for manufacturing rights

1948 Haloid and Battelle rename the process "xerography"

1950 Haloid produces the Model A copier, but it is slow and unreliable

1958 Haloid becomes Haloid-Xerox

1959 Haloid-Xerox produces the 914 Copier, the first commercially successful Xerox machine, or photocopier

1961 Haloid-Xerox becomes Xerox Corporation

1968 Carlson dies on September 19 in Rochester, New York, aged 62

Below left A demonstration of one of the more unorthodox uses of the office copier **Below right** Chester Carlson with the first model of his invention, the Xerox copier

Did you know?

An American advertisement for the Haloid-Xerox 914 Copier featured a secretary saying: "I can't type, I don't take dictation, I can't file. My boss calls me indispensable. I push the button on the Xerox 914."

The cash register was invented in 1879 by bar owner James Ritty. Its origins lie in light-fingered bar staff, a stress-relieving European vacation, and a maritime invention that was designed to count the revolutions of a ship's propeller.

CASH REGISTER

During the 1870s James Ritty (U.S.) realized that he had a problem. He was running a bar in his home town of Dayton, Ohio, but although the bar was popular and usually busy, Ritty was not making the level of profit he expected. He suspected his bartenders of stealing from him, an easy thing to do given that cash was kept in open boxes and sales were listed (if at all) in an account book by the bartenders themselves. Depressed by this, in 1878 Ritty decided to take a holiday to Europe. He visited the engine room of the steamship en route, noticed a machine that counted the revolutions of the propeller shaft, and immediately began thinking about how he could use this idea to record sales in his bar. He became so enthusiastic about the idea that he curtailed his holiday and returned home to design a cash register with his brother John.

Their first machine (pictured right), patented the following year, comprised a circular dial and a keyboard; bartenders entered the amount of each sale and the dial displayed a running total, which the manager could check at the end of the day. Crucially, the "Incorruptible Cashier," as it was known, had a loud bell that alerted the manager every time a sale was made. In 1883 James Ritty and John Birch (U.S.) patented an improved version, similar to a modern cash register, that had tabs that sprang up to indicate the amount, rather than a dial.

Having succeeded in inventing the cash register, Ritty failed to make a commercial success of it—until one of his few customers, local businessman John H. Patterson (U.S.), was so impressed with the effect the registers had on his own business that he bought a controlling stake in Ritty's company. Patterson hadn't realized that Ritty's company was running at a loss, but the deal was done. Determined to make a success of it, Patterson renamed the company the National Cash Register Co. (now better known as NCR Corp.) and went on to become a multimillionaire.

Below The first cash register, created in 1879

YOUR CHANGE 1.82 A

Below "...see how it also *figures our change!*" Advertisement for the National Cash Register Co. in *The Saturday Evening Post* (July 30, 1955)

The inventor: James Ritty

1836 Born James Jacob Ritty in Dayton, Ohio, the son of French immigrants

1870s Runs several bars in Dayton, Ohio

1878 Takes a holiday to Europe; while on board ship, he sees a machine counting revolutions of the propeller

1879 With his brother John, patents the "cash register and indicator" in the names of J. & J. Ritty (filed March 26, granted November 4)

1883 With John Birch (U.S.), patents an improved cash register similar to the modern version

1884 John H. Patterson (U.S.) buys controlling stake in Ritty's company; renames it National Cash Register Co. (NCR)

1918 Ritty dies

1922 Patterson dies

1952 NCR acquires Computer Research Corporation (CRC), of Hawthorne, California

1953 NCR establishes an Electronics Division to develop electronic applications for business machines

1974 NCR changes its name to NCR Corp.

1997 NCR Corp. signals a move from hardware only to a "full solutions provider" by acquiring Compris Technologies, Inc. and Dataworks

2000 NCR Corp. acquires Ceres Integrated Solutions and 4Front Technologies

AND ALSO...

Office safe

On November 2, 1886, Henry Brown (U.S.) received a U.S. patent for an office safe despite the fact that Alexander Fichet (France) had invented one 42 years earlier in 1844. Fichet had already patented a burglarproof lock in 1829.

Franking machine

The franking machine, or postage meter, is a common aspect of office life and was invented in 1884, surprisingly soon after the postage stamp. The idea of adhesive postage stamps was first suggested by James Chalmers (Scotland) in 1834 and introduced by the General Post Office (Britain) at the instigation of Rowland Hill (England) in 1840. The first official adhesive postage stamp, the Penny Black, was printed by Jacob Perkins, inventor of the refrigerator. Inventors soon realized that the process of buying, moistening, and affixing gummed stamps to envelopes could be simplified, particularly for high-volume users such as big businesses. In August 1884 Carle Bushe (France) filed a British patent for the invention of a machine that would print a stamp and register the amount of payment due for each stamp printed: a franking machine. Bushe's patent stated: "It is indisputable that the adoption of postage stamps did away with a great deal of trouble and annoyance, but it is impossible for progress to stop there, for that system still presents numerous inconveniences not only for the Government but for the Public. In fact, the application of adhesive stamps, which is so easy and convenient when a few only are to be used at one time, becomes a difficult matter, and entails a serious loss of time when hundreds of letters, circulars, newspapers, and so forth, have to be dispatched daily." Bushe's idea was ahead of its time, because mechanization had not reached the stage where his machines could be mass-produced cheaply enough to be a commercial success, and franking machines did not come into general use until the 1900s.

See also: Refrigerator, page 24

Carbon paper

Pellegrino Turri (Italy) is often credited with inventing carbon paper, as used in his typewriter of 1808, but in fact Ralph Wedgwood (England) patented a "carbonated paper" two years earlier in October 1806. The patent described a process involving a thin sheet of paper soaked with ink and dried with blotting paper.

See also: Typewriter, page 124

Revolving door

The revolving door was invented by Theophilus von Kannel (U.S.), whose patent was granted on August 7, 1888. This type of door was originally designed to resolve the problem of variations in air pressure in tall buildings.

Rubber stamp

On February 27, 1883, William Purvis (U.S.) received a U.S. patent for a rubber stamp, although John Leighton (England) had invented one 19 years earlier in 1864.

Elastic band

The elastic band, aka rubber band, was invented by Stephen Perry (England) and patented on March 17, 1845.

Dynamic table

In 2002 a Swiss company invented the Dynamic Meetings conference table, which, according to the company, "moves you to communicate." The chairs automatically move around the table so that delegates (or guests if used at a dinner party) can meet everyone else sitting around the table without having to move from their chairs. An electric motor in the base regulates the speed of movement, which can be set by the chairman of the meeting or host of the party. The tables are available in three models, with places for up to 12, 18, or 24 people.

Above left A masked safecracker tries to open a safe (1953)
Right The rubber band has been around since 1845

Chapter
Five

SPARE TIME

John Logie Baird is famous as the inventor of television, but Baird's mechanical system was obsolete within 15 years, and Vladimir Zworykin is considered by many to be the father of television for his invention of an electronic system.

Did you know?
Among the names suggested by prospective inventors for what became known as television were phototelegraphy, telephonoscope, radiovision, telectroscopy, audiovision, farscope, and radioscope. John Logie Baird called his scanner a televisor, which led to the word "television."

TELEVISION

Electrical engineer John Logie Baird (Scotland) suffered from chronic ill health, and moved briefly to the West Indies, hoping that the climate would be kinder to his constitution. He returned to Britain in 1922 and settled in Hastings, England, where he devised several unsuccessful inventions and began to research the idea of transmitting pictures. Short of money and still in bad health, Baird invented a mechanical television transmitter and receiver using an improved version of an optical scanning disc originally patented by Paul Nipkow (Germany). He filed a patent for his system on July 26, 1923 (granted 1924), and two years later he produced a working television from scrap materials such as tea chests, cookie tins, and darning needles. Then, in January 1926, he made the first public demonstrations of television, at the Royal Institution and at Selfridges department store on Oxford Street, London.

Meanwhile, Vladimir Zworykin (Russia–U.S.) and Philo T. Farnsworth (U.S.) were working independently on electronic television systems. Farnsworth was the first to demonstrate electronic television, based around his patented "image dissector," but it was Zworykin's cathode ray transmitter and receiver that led to the development of modern television. Just as Baird's system was based on an earlier invention, so was Zworykin's—the cathode ray oscilloscope, invented by Ferdinand Braun (Germany) in 1897. In 1923 Zworykin invented and filed a patent for a means of using a cathode ray tube as a television transmitter (granted 1938), and in 1924 he invented a cathode ray receiver: the two essential components of modern television.

At first, Baird's simpler mechanical system held sway, yielding quicker advances than electronic apparatus, but in 1936 the BBC tested the two systems against each other with alternating weekly broadcasts. Within four months, in February 1937, mechanical television was abandoned in favor of an electronic system based on Zworykin's inventions.

See also: CD/DVD, page 160

The inventors: John Logie Baird & Vladimir Zworykin

1888 John Logie Baird born on August 13 in Helensburgh, Scotland. He later graduates in electrical engineering from the Royal Technical College in Glasgow (now the University of Strathclyde)

1889 Zworykin, born Vladimir Kosma Zworykin on July 30 in Mourom, Russia. He later graduates in electrical engineering from St Petersburg Institute of Technology and then studies at the College de France, Paris

1919 Zworykin immigrates to the U.S. (naturalized a US citizen in 1924), where he works briefly for Westinghouse Electric Company

1922 Baird moves to Hastings, England, where he begins to research the idea of transmitting pictures

1923 Baird files a patent on July 26 for a: "System of transmitting views, portraits, and scenes by telegraphy or wireless telegraphy" (granted 1924). Zworykin files a patent on December 29 for a cathode ray television transmitter that he calls an iconoscope (granted December 20, 1938, after failed legal action from Philo T. Farnsworth [U.S.])

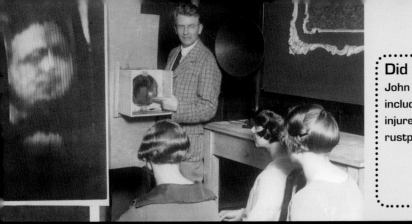

Above John Logie Baird demonstrating his televisor, with the transmitted image to the left of the picture

1924 Zworykin files a patent for a cathode ray television receiver that he calls a kinescope

1926 Baird makes the world's first public demonstration of television. Invents and patents the first video recorder, the Phonovisor, which records onto wax gramophone-type discs. Patents: "A method of viewing an object, by projecting upon it electromagnetic waves of short wavelength"—a form of radio detection that predates the invention of radar. Zworykin gains a Ph.D. from the University of Pittsburgh *See also: Radar, page 110*

1928 Baird is granted a patent for "production of optical images": the first patent for a practical application of fiber optics. Baird makes the first demonstration of color television and the first transatlantic television transmission

1937 Baird's mechanical system is abandoned by the BBC in favor of electronic

apparatus developed by Isaac Shoenberg (Russia) for EMI, based on the technology pioneered by Zworykin

1946 Baird dies on June 14 in Bexhill, England, aged 57

1982 Zworykin dies on July 29 in Princeton, New Jersey, aged 92, with more than 80 patents to his name

Above Baird's primitive televisor. The optical scanner (comprising a number of lenses mounted on a cardboard disk made from a hat box) was attached by a spindle formed from a darning needle to a motor mounted on a tea chest (1926)

Radio waves are not in themselves an invention, being a naturally occurring phenomenon, but the development of apparatus for transmitting and receiving intelligible radio waves was an invention of Nobel Prize-winning importance.

Left Guglielmo Marconi's first tuned transmitter
Below Marconi (left, wearing medal) and Pope Pius XI (right) inaugurate a radio-telephone system at the Vatican Palace for communication with the papal villa at Castel Gandolfo, 20 miles away (February 1933)

RADIO

Guglielmo Marconi (Italy) is hailed almost universally as the father of radio, but there are other claimants, including Nikola Tesla (Croatia–U.S.) and Oliver Lodge (England), who both filed earlier patents relating to wireless telegraphy, and Alexander Stepanovitch Popov (Russia), who made an intelligible radio transmission earlier than Marconi, but did not patent his invention. Whatever the claims of these and others, there is no doubt that it was Marconi who had the vision and drive to transform radio waves into a commercially practical form of communication.

Popov, Lodge, and Marconi were all working to improve on a device called a Branly tube (later named a "coherer" by Lodge), which had first been used to detect radio waves, in 1890, by Edouard Branly (France). The first step toward making the transmitter-coherer into a practical device for communication was to increase its range, which Popov and Marconi both achieved, independently, in 1895 with the invention of the radio antenna. The following year Popov succeeded in transmitting the name "Heinrich Hertz" (the man who proved the existence of radio waves) in Morse code—the first intelligible radio transmission.

Meanwhile, Marconi was determined to develop his technology and, having met with a total lack of interest from the Italian government, traveled to England, where he patented his radio apparatus in 1896. The patent described "a system of telegraphy using Hertzian waves" by which "electrical actions or manifestations are transmitted through the air, earth or water by means of electrical oscillations at high frequency." Inventions improving radio followed thick and fast, including several patented by Marconi in 1900. Between them, Lee De Forest and Edwin Howard Armstrong (both U.S.) invented a number of valves and circuits that advanced radio technology beyond all recognition, and the medium took another huge leap forward in 1947 with the invention of the transistor.

See also: AC induction motor, page 232; Clockwork radio, page 156; Transistor, page 244

The inventor: Guglielmo Marconi

1874 Born on April 25 in Bologna, Italy, to an Italian father and an Irish mother, he is later educated in Bedford, England, then at the Institute Cavallero in Florence, Italy, and at the Leghorn Technical Institute in Livorno, Italy

1894 Reads about the work of Heinrich Hertz (Germany) on electromagnetic (radio) waves

and begins his own experiments at the family's country home, Villa Griffone, Pontecchio, near Bologna

1895 Makes the inventive leap of combining the transmitter-coherer with an antenna he had been using to detect electrical storms, increasing the range of the device tenfold

Did you know?

The hypothetical existence of radio waves was first described in 1865 by physicist James Clerk Maxwell (Scotland), who theorized that electricity could cause a remote effect through what he called "electromagnetic propagation," now better known as radio waves. Many scientists rejected Maxwell's theory until Heinrich Hertz (Germany) proved it in 1887. Having established this momentous proof, Hertz reputedly said: "I don't see any useful purpose for this mysterious, invisible electromagnetic energy."

Guglielmo Marconi's mother was Anne Marconi, née Jameson, a member of the family famous for Jameson's Irish whiskey.

1896 Travels to England, where he files a patent for: "Improvements in transmitting electrical impulses and signals, and in apparatus therefor" (granted June 2, U.S. patent granted July 13, 1897, and reissued on June 4, 1901)

1897 Establishes the Wireless Telegraph & Signal Co. Ltd. to develop the technology

1899 Establishes the Marconi Wireless Telegraph Co. of America for further development

1900 Files a patent on April 26 for improvements to his radio apparatus (granted the same year)

1901 Makes the world's first transatlantic radio transmission on December 12: three dots representing the letter *S*, sent from Cornwall, England, to Newfoundland, Canada

1909 Shares the Nobel Prize for Physics with Karl Ferdinand Braun (Germany) for their "contributions to the development of wireless telegraphy"

1919 Appointed Italian delegate to the Paris Peace Conference following the First World War

1937 Dies on July 20 in Rome, aged 63, having filed more than 100 patents relating to wireless communication

Neither clockwork nor radio was a new invention in 1991, but combining the two technologies to invent the clockwork radio in that year shot Trevor Baylis to fame and vastly improved life for thousands of people in Africa.

CLOCKWORK RADIO

Early in 1991 inventor Trevor Baylis (England) saw a television documentary about the spread of AIDS in Africa, which asserted that the death toll would be far lower if people were better informed about the disease and about safe sex. But the dissemination of such information to remote villages was almost impossible because there were no televisions, and radios were rarely used because there was no electricity supply and batteries were expensive.

Baylis, who has spent much of his career inventing devices to help the disadvantaged, was moved to do something about the situation. Instead of inventing an alternative means of communication, he invented a way of powering radios in places with no electricity: "I found a very small DC electric motor, which I knew when run in reverse would generate electricity as a dynamo, and I put that into the chuck of an ordinary hand brace [drill] and then stuck it into a bench vice. I held the little motor, having joined the two wires to a cheap transistor radio, and then, as I turned the handle, I got the first bark of sound from that radio. So I knew then I was on my way, that 'Eureka moment' if you like." Months of development followed the Eureka moment before Baylis filed a patent for a geared clockwork system that could power a radio for 14 minutes after 2 minutes' winding.

The invention was inspired *by* broadcast information, its purpose was *to* broadcast information, and a manufacturer was eventually found *through* broadcast information. Christopher Staines and Rory Stear (both South Africa) saw Baylis demonstrating his invention on the television program *Tomorrow's World* and helped him to establish BayGen Power Industries to manufacture the radio. BayGen changed not only the lives of villagers who now had access to broadcast information through the radio, but also the lives of hundreds of disadvantaged workers who benefitted from a policy of recruiting disabled people to work in BayGen's factory.

See also: Radio, page 154

Did you know?

One of the problems in developing the clockwork radio was in producing a mechanism small enough, but when BayGen tested reactions to the radio in Africa, the response was that it could be bigger. Unlike Europeans, for whom miniaturization is desirable, the Africans saw the size of the radio as a measure of their status, removing at a stroke the problem of downsizing the mechanism.

More recently Trevor Baylis has applied the wind-up technology to laptop computers, potentially opening up an even greater web of information to remote villages.

The inventor: Trevor Baylis

1937 Born Trevor Graham Baylis in Kilburn, London, he later trains in mechanical engineering and physical education

1991 Invents the clockwork radio and files a patent on November 19

1995 With Christopher Staines, Rory Stear, and the Liberty Life Group (all South Africa), Baylis establishes BayGen Power Industries to produce the clockwork radio. BayGen Products Pty. begins producing the clockwork radio in Cape Town, South Africa, under the trademark Freeplay

1996 The Freeplay wins the BBC Design Awards for Best Product and Best Design. Baylis's patent is rescinded after the citation of five prior patents, the earliest dating from 1923

1997 Baylis produces the Freeplay 2, which will run for 1 hour after just 30 seconds' winding. Appointed an OBE (Officer of the Order of the British Empire) and awarded the President's Medal of the Institute of Mechanical Engineers

Opposite top Trevor Baylis, inventor of the wind-up radio, in his workshop (February 1999) **Below** Bushman Jakob Malgas listens to a BayGen clockwork radio in the Kalahari Gemsbok National Park, South Africa

Did you know?

As well as being an inventor, Trevor Baylis swam competitively for Great Britain, served as an army PT instructor, performed with the Berlin State Circus, and worked as an underwater stuntman and escapologist.

The word "broadcast," as used for radio and television programs, derives from an 18th-century agricultural term referring to a method of sowing seeds by scattering them in all directions by hand.

The phonograph and the kinetoscope were soon obsolete, but as the precursors of the record and cinema industries they were both enormously important inventions—two of nearly 1,100 patented by the nonpareil of inventors, Thomas Edison.

PHONOGRAPH/KINETOSCOPE

Thomas Alva Edison (U.S.) was the most prolific inventor of all time, with 1,093 U.S. patents to his name. Widely known as The Wizard of Menlo Park for his undoubted inventive genius, he was also notorious as an uncompromising man who ruffled the feathers of colleagues such as fellow inventor and engineer Nikola Tesla (Croatia–U.S.), and whose ideas were sometimes misguided—as when he doggedly championed direct current (DC) electricity over alternating current (AC). Edison's best known invention is the incandescent light bulb (1879), but two of his other inventions, the phonograph and the kinetoscope, shaped two of the biggest leisure industries of the 20th century—the music business and the film industry.

In 1877 Edison invented what he patented as a "Phonograph or Speaking Machine," a device that recorded sound using a stylus that made indentations on a tinfoil cylinder, and played them back with a second stylus that retraced the indentations. The phonograph was rendered obsolete just ten years later, when Emile Berliner (Germany–U.S.) invented the flat-disc gramophone as an improvement on Edison's cylinder machine, but, nonetheless, Edison had set in motion an entire industry based around the recording of music and sound. Ironically, music was not one of the potential uses listed in Edison's patent—top of the list was dictation.

Fourteen years later, in 1891, Edison and his employee, William Kennedy Laurie Dickson (U.S.), invented a motion–picture camera known as the kinetograph and a peephole viewer for the resulting pictures, which they called the kinetoscope. In 1894 a kinetoscope parlor in New York provided the first commercial viewings of a motion picture, but, like the phonograph, the kinetoscope proved short-lived. However, it did inspire Louis Lumière (France) to build his *cinématographe,* with which he made the world's first public screening of a motion picture (as opposed to peephole viewing) in 1895, marking the birth of cinema.

See also: AC induction motor, page 232; Air brake, page 88; CD/DVD, page 160; Pinball, page 178; Light bulb, page 10

The inventor: Thomas Alva Edison

1847 Born on February 11 in Milan, Ohio, the son of a Canadian roof-tile maker, he grows up with little formal education or training

1868 Files his first patent on October 28, for an "electrographic vote recorder" (granted June 1, 1869)

1871 Invents the ticker-tape automatic repeater for stock exchange prices. Establishes an industrial research laboratory in Newark, New Jersey (often described as the world's first research and development laboratory)

1876 Moves his laboratory to Menlo Park, New Jersey. Invents and patents a method of stencil duplication, which he calls autographic printing

1877 Invents and files patents for the "Phonograph or Speaking Machine" (U.K. patent granted 1877, U.S. patent granted 1878). Invents a carbon telephone transmitter (microphone)

1879 Invents and files a patent for an incandescent light bulb (patent granted 1880, rescinded 1883)

1880 Produces the world's first talking doll, using a recording of a human voice saying: "Mommy, Daddy." Discovers the principle of thermionic emission (initially known as the Edison Effect), vital to the development of thermionic valves for use in radio. Invents and files a patent for a voltage indicator (granted 1883 as the first patent for an electronic device)

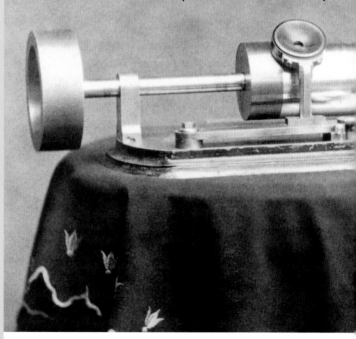

Did you know?

On waiting, Edison said: "Everything comes to him who hustles while he waits." But Edison did not like waiting —he also said: "I am long on ideas, but short on time. I expect to live only about a hundred years." On success, he wrote: "You've got to make the damn thing work....I have failed my way to success," and, later: "I know several thousand things that won't work."

Below Thomas Edison and his speaking phonograph (April 1878)
Right Edison kinetoscope (1894)

1882 Granted 75 patents, his most prolific year

1891 Invents and files a patent for a motion picture camera, the kinetograph, and a peephole viewer, the kinetoscope (patent granted 1893)

1893 Builds the world's first motion-picture studio

1908 Establishes the Motion Picture Patents Company and begins buying up motion-picture patents to create a monopoly (dissolved in 1917 for breaching antitrust laws)

1912 Offered the Nobel Prize for Physics jointly with Nikola Tesla (Croatia–U.S.), but Tesla refuses to be associated with Edison, so the prize is awarded instead to physicist and

inventor Nils Gustav Dalén (Sweden)

1914 Invents the alkaline storage battery

1927 Elected a member of the National Academy of Sciences

1928 Awarded the Congressional Gold Medal

1931 Files his last patent on January 9, for: "Holder for Article to be Electroplated" (granted May 16, 1933). Dies on October 18 in West Orange, New Jersey, aged 84, with 1,089 U.S. patents to his name. Four more are awarded posthumously, setting a record of at least one patent per year for 65 consecutive years

The videodisc was invented by John Logie Baird as early as 1926, but the idea did not become a commercial success until 1996 with the introduction of the digital versatile disc (DVD), a development of the earlier compact disc (CD).

CD/DVD

Above Austrian conductor Herbert von Karajan conducts a rehearsal (c. 1950
Right CDs hold 74 minutes of music: the length of Beethoven's Ninth Symphon

The inventor of television was also the inventor of the first video recorder. In 1926 John Logie Baird (Scotland) was granted a patent for a device that recorded visual images on a 10-inch wax disc, using the same principle as the gramophone. The idea of recording visual images on disc was reinvented 46 years later by Philips Electronics (Netherlands) as Philips Laservision, which they demonstrated in 1972 and launched in the U.S. in 1980 and in Europe in 1982—the same year as the commercial launch of the CD, which had been developed by Philips in conjunction with Sony (Japan).

Both the CD and the laser-read videodisc originated from the invention of an optical disc by James T. Russell (U.S.). Russell was frustrated by the wear and tear of styli on vinyl records and, in 1965, patented the idea of storing information on a disc to be read optically by a laser. Philips developed Russell's concept to store video images, and from 1969 Klaas Compaan and Piet Kramer (both Netherlands) developed the Video Long Player: a 12-inch videodisc that was first demonstrated in 1972 and eventually launched in 1980 as Laservision. Meanwhile, from 1975 Philips's engineer Lou Ottens (Netherlands), director of audio R&D, assigned a team to study the possibilities of hifi-audio on a small optical disc. After joint development with Sony, the invention was launched in 1982 as the now familiar 4.8-inch compact disc, the use of which has since spread from music to computers, in the form of the CD-ROM.

Baird's disc and Philips's Laservision had been ahead of their time: The idea was right, but the product was impractical. However, advances in disc and laser technology in the 1990s led to the development by Philips, Sony, Matsushita (Japan), and Toshiba (Japan) of the digital versatile disc. A DVD could store a vast amount of information on a disc the same size as a CD— easily enough information for a feature film, which meant that DVD almost immediately superseded videotape as the medium of choice for recording visual images.

See also: Television, page 152

Discs time line

1877 Thomas Alva Edison (U.S.) invents the phonograph, the first machine to record and reproduce sound (it does so on a foil cylinder) *See also: Phonograph/Kinetoscope, page 158*

1887 Emile Berliner (Germany— U.S.) invents the gramophone, the first machine to record sound onto a disc

1926 John Logie Baird (Scotland) invents the video recorder, which records onto a gramophone-type wax disc

1931 RCA Victor (U.S.) produces the first $33^1/_3$ rpm long-playing audio disc

1948 Columbia (U.S.) produces the first vinyl long-playing audio disc (developed by Peter Goldmark [U.S.]) and coins the term "LP"

1972 Philips (Netherlands) demonstrates the laser-read video long player (*see* 1980 below)

1975 Decca (Britain) and AEG (Germany) launch the Teldec videodisc (not a great commercial success)

1980 Philips and Sony (Japan) announce the invention of the audio compact disc (CD). Philips launches the VLP as Philips Laservision, the first videodisc to be read by laser. RCA launches the Selectavision videodisc system

1982 The first CD player, the Philips CD100, goes on sale in Japan (CD players are first sold in Europe the following year)

c. 1985 CD-ROM introduced

1995 Standard DVD format agreed between manufacturers

1996 Specifications are published for DVD-ROM and DVD-video. The first DVD-video players go on sale in Tokyo, Japan

1997 DVD is launched in North America

1998 DVD is launched in Europe

Did you know?

The size of the prototype CD was increased from 4.6-inch to 4.8-inch to increase playing time to 74 minutes, to comply with the requirement of Sony chief executive Akio Morita (Japan) that a CD must be able to hold the whole of Beethoven's Ninth Symphony. Morita was a personal friend of Herbert von Karajan (Austria), who conducted the Berlin Philharmonic in what is considered by many to be the definitive version of the symphony. Karajan declared that, compared with CDs: "All else is gaslight."

The "anti-jog" device on portable and in-car CD-players does not physically prevent the laser scanner from skipping if the machine is jogged. Instead, there is an electronic delay between scanning and playback, which means that if the machine is jogged, it can play back what it has "remembered" while it refinds its place to continue scanning. Philips demonstrates this with a player that continues playing even if the CD is taken out of the machine.

The first CD to sell a million copies was *Brothers in Arms* by Dire Straits (1986).

George Eastman did not invent photography, but he did bring it within reach of everyone with his invention of the first roll-film camera, the Kodak No.1—the most famous, but not the first, of his many photographic inventions.

ROLL-FILM CAMERA

Until George Eastman (U.S.) invented the roll-film camera in 1888, photography was the preserve of experts, something that involved specialist equipment and required the knowledge to develop and print one's own film. The roll-film camera was one of several inventions by Eastman aimed at making photography simpler, beginning in 1879 with the invention of a machine for coating dry photographic plates. Previously, photographers had used wet plates, which required chemicals to coat the plate immediately before taking the photograph, but Eastman's dry plates were precoated, saving time and reducing the complexity of the operation.

Having set up a company to manufacture dry plates, Eastman realized that it made good business sense to turn photography into a mass-market activity, so he began to devise ways of making the process simpler still. In 1884 he invented a flexible paper film as a replacement for glass plates, and in 1885, with William H. Walker (U.S.), he invented a roll holder for his film that could be fitted to an existing plate camera. Then came the invention that was to revolutionize photography: the Kodak No.1. It was the first camera to have a built-in roll-holder, and was specifically designed for ease of use, with a string to open the shutter, a button to release it again and a key to wind on the film. When the 100-exposure film was finished, the user sent it to the Kodak factory, where the film was removed and processed before the camera was returned, reloaded, and ready to be used again. The advertising slogan said it all: "You press the button—we do the rest."

For the rest of his life Eastman continued patenting inventions that popularized photography and earned him enough money to become a renowned philanthropist—he donated over one million dollars to arts, education, and medical and scientific research. Then, on March 14, 1932, after being diagnosed with a spinal disorder, he shot himself. He left a note reading: "To my friends. My work is done. Why wait?"

The inventor: George Eastman

1854 Born on July 12 in Waterville, New York, the son of a teacher

1868 Leaves school aged 14 to support the family after the death of his father in 1862

1874 Becomes a bookkeeper at Rochester Savings Bank, income from which enables him to pursue his hobby of photography

1877 Reads about a dry-plate process invented in England

1879 Invents and patents a machine for coating dry plates (patents granted 1880)

1880 Forms a partnership with Henry A. Strong (U.S.) and begins manufacturing dry plates

1881 Establishes the Eastman Dry Plate Co. (later Eastman Dry Plate & Film Co., later Eastman Co., later Eastman Kodak Co., later Kodak)

1884 Invents and patents a flexible, paper-backed photographic film known as stripping film because it has to be stripped off the backing

1885 With William H. Walker (U.S.), invents and patents a mahogany film roll holder

1888 Invents and patents the Kodak No.1, the first roll-film camera

1889 Employs Henry M. Reichenbach (U.S.) to invent a paperless flexible film, which Reichenbach patents on behalf of Eastman Dry Plate & Film

Co. and which becomes the first celluloid roll film to be produced commercially. Unfortunately, Hannibal Goodwin (U.S.) has already patented a similar film (in 1887), which in 1914 is given precedence after nearly 20 years of legal battles, resulting in Eastman paying hefty retroactive royalties

1891 Launches daylight loading film, enabling users to load cameras themselves instead of having to send them to the factory

1895 Launches the Pocket Kodak camera

1898 Launches the Folding Pocket Kodak camera (with a bellows to make it collapsible)

1900 Launches the Kodak Brownie camera, designed by Frank Brownell (U.S., b. Canada) and named not after him but after the fairy-like "Brownies" of children's writer and illustrator Pamela Cox (Canada)

1923–32 Launches various 0.6 in. (16mm) and 0.3 in. (8 mm) cine cameras, film and projectors

1932 Kills himself on March 14, aged 77

Did you know?

The name Kodak was originally intended simply as the name of George Eastman's 1888 camera, and only became part of the company name in 1892, the same year that the famous yellow packaging was adopted. Eastman said: "I devised the name myself.... The letter 'K' had been a favorite with me—it seems such a strong, incisive sort of letter....It became a question of trying out a great number of combinations of letters that made words, starting and ending in 'K.' The word Kodak was the result."

**Opposite
top** Proud owners of Kodak cameras (c. 1900)
Above Photograph by painter Frederick Church of George Eastman on the deck of the *Gallia*, with a Kodak box camera in his hand (1890)

In 1943 a three-year-old girl asked a very simple question: Why couldn't she look at a photograph of herself as soon as it was taken? Her father, Edwin Land, went away and invented a camera and film that would allow her to do just that.

Did you know?

In terms of picture quality, instant photography is usually considered the poor relation of conventional photography, but in 1982 David Hockney (England) turned it into an art form, producing a series of photo-collages with the collective title *Cameraworks*.

POLAROID CAMERA

Jennifer Land (U.S.) provides the perfect example of how a naïve approach to a situation can lead to a breakthrough that a trained mind would never consider. Only a child (or an inventor) would have thought to ask why a photograph could not be seen the instant it was taken, and only an inventor would bother to find an answer rather than dismissing the question. Jennifer's father, Edwin Herbert Land (U.S.), said later: "Within an hour the camera, the film and the physical chemistry became so clear that with a great sense of excitement I hurried to a place where a friend was staying to describe to him in detail a dry camera that would give a picture immediately after exposure."

The concept may have been as instantaneous as the photography that was to follow, but the development of the idea was not quite as rapid. It took three years to perfect the camera, which worked by passing the exposed negative between a set of rollers that ruptured a container of chemicals that, in turn, developed and printed a positive image. The camera then spat out a sepia-toned (brown and white) print that could be peeled away from the attached negative less than a minute after exposure. Land demonstrated his invention to the American Optical Society on February 21, 1947, and launched the Polaroid Land Instant Camera on November 26, 1948, to immediate acclaim, with sales in the region of five million dollars in the first year.

Land did not rest on his laurels, but immediately began making improvements to his instant camera, filing some 300 patents over the next decade. The first improvement was the production of black and white rather than sepia prints, then came instant color prints in 1963, and then, in 1972, Polaroid produced the SX-70 camera, whose photographs did not require peeling from their negative. The new SX-70 allowed users to watch prints develop in daylight before their very eyes, but even that remarkable advance seems rather quaint now, in the age of truly instantaneous digital photography.

The inventor: Edwin Herbert Land

1909 Born on May 7 in Bridgeport, Connecticut, the son of a scrap metal merchant, he later studies at Norwich Academy before embarking on a physics degree at Harvard University

1932 Invents a light-polarizing filter that he calls Polaroid, the first synthetic polarizer of light, and files a patent for: "Polarizer and method of making the same" (U.K. patent granted 1934). Leaves Harvard and founds Land-Wheelwright Laboratories with George Wheelwright III (U.S.)

1935 Land-Wheelwright begins making polarizing filters for Eastman Kodak (U.S.)

1936 The American Optical Company begins using Polaroid in sunglasses

1937 Land establishes the Polaroid Corporation

1942–45 Invents a number of weapons for the National Defense Research Council

1943–46 Invents and develops the instant camera

1948 Files a patent on February 3 for: "Photographic Apparatus"—the Polaroid Land Instant Camera (granted February 27, 1951). Files a patent on December 11 for: "A photographic product comprising a rupturable container carrying a photographic processing liquid"—instant film (granted February 27, 1951)

1949–60 Polaroid Corp. is granted some 300 patents, 120 of them in Land's name

1963 Polaroid Corp. introduces Polarcolor film, invented by Polaroid chemist Howard G. Rogers (U.S.) at Land's behest

1972 Polaroid Corp. introduces the SX-70, the first instant camera to produce photographs that do not require peeling from their negative

1980 Land founds and finances the Rowland Institute for Science

1982 Retires from Polaroid Corp.

1991 Dies on March 1 in Cambridge, Massachusetts, aged 81, having been awarded 14 honorary degrees and 37 medals and awards, and with 533 patents to his name

Below Inventor Edwin Land in his office in Cambridge, Massachusetts, holding a Polaroid portrait of himself posing, moments earlier, with the SX-70 Polaroid camera (November 1972) **Right** The first commercially produced Polaroid camera, the Model 95, first manufactured in 1948

The origins of tennis lie in an 11th-century French game called *jeu de paume.* This developed into "real" or "royal" tennis, and modern lawn tennis evolved from a variation invented by Maj. Walter Clopton Wingfield in 1873.

TENNIS

Above French illustration of aerial tennis, entitled "In the Year 2000" (1899) **Opposite** Walter Wingfield's hourglass-shaped lawn tennis court in action

Tennis as we know it evolved from a game first played in the cloistered courtyards of French monasteries c. 1050. In England this then developed into "real" tennis, for which special brick-built indoor courts replaced the monastic courtyards. The earliest record of an outdoor version of the game is a reference to field tennis in *Sporting Magazine* (England) on September 29, 1793, but the modern form of lawn tennis evolved from a game called *sphairistike,* devised in 1873 by Maj. Walter Clopton Wingfield (Wales), and from the invention of what Wingfield patented in 1874 as: "A New and Improved Portable Court for Playing the Ancient Game of Tennis."

Wingfield first played *sphairistike* in 1873 with a group of friends in Wales, and published details of the game in a pamphlet entitled: *The Game of Sphairistike: or Lawn Tennis*—not surprisingly, it was the alternative name that stuck. The game was played on an hourglass-shaped court, narrower at the center than at the ends, with an H-shaped arrangement of nets forming side walls and a central division. Wingfield's patent states: "By this simple apparatus a portable court is obtained by means of which the old game of tennis, which has always been an indoor amusement, and which few can enjoy on account of the great expense of building a brick court, may be made an outdoor one, and brought within reach of all."

During 1874 Wingfield advertised his invention in newspapers and began selling boxed sets of the equipment required to play the game. Lawn tennis quickly became popular as an alternative to croquet and another new sport, badminton (first played at Badminton House, Gloucestershire, c. 1870). The first official set of rules was drawn up by the Marylebone Cricket Club (MCC) at Lord's Cricket Ground in 1875, and the first All-England Lawn Tennis Championship was played in 1877 in Wimbledon at the All-England Croquet and Lawn Tennis Club, as it was then known, which has administered the game ever since.

The inventor: Walter Clopton Wingfield

1833 Born in October in Ruabon, Wales, and later serves in the British Army as a member of the King's Dragoon Guards

1873 Invents *sphairistike.* The first game is played at Nantclwyd, Wales, on September 29

1874 Files a patent on February 23 for a portable tennis court (granted July 24). Advertises the game in newspapers from March 7

1875 First official set of rules drawn up by the MCC. Mary Ewing Outerbridge (U.S.) sees British Army officers playing Wingfield's game in Bermuda. She buys a set and introduces it to the U.S., where it is first played at the Staten Island Cricket & Baseball Club, of which her brother is secretary

1877 The first All-England Lawn Tennis Championship is held, at Wimbledon

1897 Wingfield publishes a book describing another new sport, bicycle gymkhana, but this is not as successful as lawn tennis

1900 Invents and patents a scoring board for the card game bridge

1912 Dies at Rhysnant Hall, Montgomeryshire, Wales, aged 79

Did you know?

The name "tennis" derives from the French *tenez,* meaning "hold" or "attention," said to be because the server in *jeu de paume* would shout "*Tenez!*" to warn their opponent that they were about to serve the ball.

The idea of metal rackets was conceived by tennis player and sportswear manufacturer René Lacoste (France), who patented the first steel racket in 1960.

If things had been different, lawn tennis might have evolved as ice tennis. Wingfield's patent stated: "The above-described court can be erected in a few minutes on a lawn, on ice, or in any suitable sized space either in or out of doors."

Golf itself is not an invention, but since people first hit a ball into a hole with a stick more than 2,000 years ago it has inspired an inordinate number of inventors to find ways of improving the game and the equipment used to play it.

GOLFING INVENTIONS

Golf balls evolved from wooden spheres to hand-sewn leather cases filled with feathers, known as Featheries and invented c. 1618. Featheries were superseded by solid gutta-percha balls, known as Gutties and invented by Rev. Adam Paterson (Scotland) in 1848, and these in turn were superseded by a ball comprising a solid case filled with rubber threads wound around a solid core. This type of ball, known as the Bounding Billy, was invented by Coburn Haskell and his playing partner Bertram George Work (both U.S.) and patented by Haskell in 1898. Two years later W. Millinson of the Haskell Golf Ball Co. (U.S.) patented a machine for winding the rubber thread, leading to the mass production of the new type of ball. The rubber core and winding remained standard for the greater part of the century, despite the threat from the invention (by Donald B. Poynter, U.S., in 1968) of a walking golf ball. Poynter's patent stated: "The walking golf ball comprises a spherical hollow casing containing a motor, and legs activated by the motor cause the golf ball to advance with a walking motion toward a cup or other target...." Which puts a new spin on the famous description of golf as "a good walk spoiled."

Early golf clubs ranged from sticks to shepherd's crooks and then progressed to purpose-designed wooden clubs. Arthur F. Knight (U.S.) invented the first steel-shafted golf club, patented in November 1910, but this revolutionary idea was nothing compared with the invention in 1960, by Ashley Pond III (U.S.), of a "Breakable Simulated Golf Club." Pond's patent stated that the object of his invention was "to provide a golf club for temperamental golfers wherein the shaft of the club is deliberately constructed to break when struck against the ground, a tree, or other inanimate elements when the anger of the golfer reaches a mercurial height, and wherein the emotion of the golfer requires some physical manifestation to achieve emotional release."

Did you know?

Featheries were made from hand-sewn horsehide or cowhide cases stuffed with wet goose feathers. As they dried out the feathers expanded and the leather shrank, creating an extremely hard and compact golf ball.

The first golf shots to be played on the moon were struck in February 1971 by astronaut Alan Shepard (U.S.), commander of *Apollo 14*. The Royal & Ancient Golf Club (Scotland) later sent Shepard a telegram that read: "Warmest congratulations to all of you on your great achievement and safe return. Please refer to the Rules of Golf section on etiquette, paragraph 6, quote—before leaving a bunker a player should carefully fill up all holes made by him therein, unquote."

Opposite top An inscribed block of raw leaf gutta-percha. Gutta-percha is a resin closely related to rubber, and takes its name from the Malay *gutta*, meaning gum or resin, and *percha*, the tree that produces it **Below** Golf's origins are long and storied, and well documented

Golf: some inventions

c. 1618 The Featherie is introduced

1848 Rev. Adam Paterson (Scotland) invents the solid gutta-percha ball, or Guttie

1898 Coburn Haskell & Bertram George Work (both U.S.) invent the rubber-cored golf ball, patented by Haskell

1899 George F. Grant (U.S.) invents the golf tee

1900 Haskell Golf Ball Co. invents and patents a machine for mass-producing rubber-cored golf balls

1905 William Taylor (England) invents and files a patent for the now ubiquitous dimple pattern golf ball (granted 1907). Spalding buys the U.S. rights in 1908

1910 Arthur F. Knight (U.S.) invents and patents the steel-shafted golf club

1914 Charles Theophilus Ramsay (England) invents and patents: "An Improved Optical Instrument for use in Playing the Game of Golf or like Ball Games." The patent states that the invention "may be in the form of pince-nez or spectacles" and has "a species of opaque blinkers" in order to force players to keep their eye on the ball

1960 Ashley Pond III (U.S.) invents a breakable simulated golf club for which he files a patent on November 25 (granted April 30, 1963)

1967 Arthur Pedrick (England) invents a golf ball with flaps to assist its flight and with a metal core to allow golfers to find lost balls using a Geiger counter (patent filed July 18, 1967, granted July 31, 1968) *See also: Arthur Pedrick, page 272; Geiger counter, page 242*

1968 Donald B. Poynter (U.S.) invents a walking golf ball for which he files a patent on July 17 (granted March 30, 1971)

1970 Arthur Pedrick (England) receives a patent for a golf tee incorporating a magic eye to monitor the path of the club as it approaches the tee. If the magic eye senses the club will hook or slice the ball, it triggers a blast of compressed air to blow the ball off the tee, saving the player from a mis-hit and having to search for a lost ball *See also: Magic eye, page 234*

2001 John Carr (U.S.) receives a patent for an: "Aquatic golf swing training device and method for enhancing golf swing memory and strength." The invention, which is basically a paddle on a stick to provide resistance when used in the water, is described in the patent as: "A manually grippable handle secured to a shaft that has integrally secured thereto adjacent [to] the handle a hydrodynamically adjustable paddle that may be manually physically altered to provide a variable resistance to a user that grips the handle and swings the golf swing training device through water that compresses the aquatic environment"

In 1949 Danish carpenter Ole Kirk Christiansen invented what he called "automatic binding bricks." Then, 9 years later, his son Godtfred invented the stud-and-tube coupling system that now instantly identifies LEGO building bricks.

LEGO timeline

1916 Carpenter Ole Kirk Christiansen establishes a business in Billund, Denmark, selling wooden stepladders and ironing boards

1932 Christiansen adds wooden toys to the company's repertoire of products

1934 Christiansen names his toys Lego, from the Danish *leg godt*, meaning "play well"

1947 Christiansen installs Denmark's first injection moulding machine and begins manufacturing plastic as well as wooden toys

LEGO

Several fairy tales involve a poor carpenter who, through hard work and perseverance, makes his fortune and lives happily ever after, and the story of LEGO is no exception. Carpenter Ole Kirk Christiansen (Denmark) was 25 years old when he set up business in his home town of Billund, Denmark, making wooden stepladders and ironing boards. But business was poor and in 1932, during the worldwide depression that followed the Wall Street Crash of 1929, Christiansen added wooden toys to the company's repertoire in order to make ends meet. Two years later he gave his toys the name Lego, from the Danish *leg godt*, meaning "play well."

In 1947 Christiansen bought Denmark's first injection moulding machine and began making plastic toys, including an invention that he called Automatic Binding Bricks: children's building blocks with studs on top and hollow undersides, enabling them to fit into each other. Christiansen began selling these blocks in 1949, but the real breakthrough came in 1958 when his son, Godtfred Kirk Christiansen, improved on the binding concept by inventing and patenting the idea of creating moulded hollow cylinders within the underside of each brick. This enabled the studs on top to interlock more firmly with the next brick, thereby making a much stronger connection—these really were "binding bricks" in a way that the originals never had been.

Godtfred's invention soon proved popular around the world, and in 1960 the company stopped making wooden toys altogether to concentrate on what by then were known as LEGO building bricks (later the LEGO System). By the end of the millennium there were more than 2,000 different LEGO components, and LEGO was estimated to be present in 80 percent of European and 70 percent of American households. Christiansen could hardly have imagined in 1932 that his line of toys would develop into something so popular, or bring happiness to so many children.

1949 Christiansen begins selling plastic "Automatic Binding Bricks," a forerunner of LEGO building blocks

1958 Christiansen's son, Godtfred Kirk Christiansen, invents and patents the stud-and-tube coupling system that now characterizes modern LEGO blocks

1960 The LEGO Group (Denmark) ceases production of wooden toys

1967 The LEGO Group introduces DUPLO, larger blocks for smaller children

1997 The LEGO Group introduces LEGO MINDSTORMS, blocks programmed to carry out various activities

Left "Keep feeding their imagination": UK advertisement from 1970s
Above The original Legoland in Jutland, Denmark

Did you know?

At the turn of the millennium the tallest LEGO structure was a tower built in the appropriately named town of Tallinn, Estonia, by a team of more than 6,000 schoolchildren. The tower comprised 391,478 eight-stud blocks and stood 24.91 meters tall. The *longest* LEGO structure was also built by schoolchildren—a 610.8 meter millipede comprising more than 1.6 million bricks, built between July 8 and 13, 2000, at LEGOland California.

Realizing that a doll for boys would never sell, Don Levine invented a new genre of toys—the action figure. GI Joe took the American market by storm, and launched a secondary attack on the British market two years later as Action Man.

GI JOE/ACTION MAN

GI Joe/Action Man time line

1963 Stan Weston (U.S.) invents the concept of a military doll and sells the idea to Hassenfeld Brothers Inc. (U.S., later Hasbro)

1963–64 Hassenfeld Brothers Inc. conceives the idea of an "action figure," develops GI

Joe, and files a patent for the figure (granted October 11, 1966)

1964 GI Joe is launched on the American market

1966 GI Joe is launched on the British market as Action Man

Early in 1963 merchandising executive Stan Weston (U.S.) conceived the idea of a military doll with a range of accessories. He discussed the idea with his friend Larry Reiner (U.S.), who said: "You have to make the figure articulated, so you can get him behind a machine gun or what-have-you"—a significant step toward the invention of GI Joe. On April 11, 1963, Weston tried to sell the idea of what he called "rugged-looking scale dolls for boys" to Don Levine (U.S.) of toy manufacturer Hassenfeld Brothers Inc. (U.S., later Hasbro). The phrase "dolls for boys" almost killed the idea, but Levine realized that it would sell if it were never referred to as a doll. He decreed that the concept would be referred to as a "moveable fighting man" or "action figure," and after a year of frantic development led by Levine, GI Joe was launched in 1964 to huge acclaim. The following year Palitoy (England) secured a licensing deal, and on January 30, 1966, GI Joe was launched as Action Man in Britain, where he became a cultural icon of the 1970s.

The Hasbro and Palitoy design teams immediately began inventing new features, the first of which was a talking figure. This was inspired by an advertisement for GI Joe with a voice-over shouting, "Storm the hill, men!" which prompted Levine to say to Hassenfeld: "Wouldn't it be great if we could get our guy to talk like that?" Next came realistic hair, invented by Bill Pugh (England), who had seen an item about "flocking" (spraying nylon fibers onto walls for acoustic purposes) on the science program *Tomorrow's World* and decided to apply the process to Action Man's head. In 1972 Pugh had another brainstorm, this time gripping hands molded from a new type of plastic called Kraton, and in 1976 Hasbro invented moving "Eagle Eyes," introduced to Action Man the following year. GI Joe met his demise in 1977 and Action Man in 1984, but both were relaunched in the 1990s to renewed acclaim.

and later across Europe under various names. Don Levine (U.S.) conceives the idea of Talking Joe

1968 Talking Joe is launched in Britain as Action Man Talking Commander

1969 Bill Pugh (England) conceives the idea of

realistic hair (introduced in 1970)

1972 Pugh conceives the idea of gripping hands (introduced in 1973)

1976 Hasbro conceives and introduces Eagle Eyes (introduced to Action Man in 1977)

1977 12 in. (31 cm) G.I. Joe figures replaced temporarily by 3.75 in. (9.5 cm) figures

1978 Pugh's team introduces an all-new Dynamic Physique for Action Man (designed and patented 1977)

1984 Production of Action Man ceases

1991 Hasbro launches a special limited edition of 12 in. (31 cm) GI Joe figures, which is so successful that it leads to a full-scale relaunch of GI Joe

1993 Hasbro relaunches Action Man in Britain and Europe

Opposite A U.S. soldier shows off his GI Joe during a break from military operations in Afghanistan (February, 2003) **Above** A barroom brawl breaks out at the GI Joe convention

Did you know?

Palitoy's former marketing manager Nick Farmer (England) was not convinced that flock hair would be durable enough for a toy of Action Man's rugged nature, so he devised a unique quality-control test: "I used to drive a Mini at that time so I got one of these brand new heads and filled it up with resin car body filler;

I drilled it out and tapped in a thread and made a gear knob out of it, so the gearstick on my Mini had one of these Action Man heads with lifelike hair. It was a pretty severe test because you're constantly changing gear if you're in a Mini. It did 50,000 miles and there was no loss of hair at all, so I had to eat my

words on that one and say that Bill Pugh had got it exactly right."

Apart from his numerous inventions relating to Action Man, Bill Pugh's most widely recognized invention was the plastic lemon, still used today as a container for Jif lemon juice.

Monopoly became popular as cheap entertainment during the Depression of the 1930s, but it has since come to mean big bucks—at the turn of the millennium it was the best-selling and most widely played board game in the world.

174 MONOPOLY

On August 31, 1935, Charles B. Darrow (U.S.) filed a patent for what was to become the world's best-selling board game, but the origins of Monopoly go back much further than that. As early as 1903, political radical Lizzie J. Magie (U.S.) filed a U.S. patent (granted 1904) for an anti-capitalist board game that she called The Landlord's Game, which featured property squares with rental prices, as well as railways, utilities, and a jail. The game evolved through various homemade versions (and some published variations) until Darrow popularized the now familiar form of Monopoly, reversing its political stance to make it a pro-capitalist game. Darrow may not have invented Monopoly, but he certainly provided the impetus that would make it a worldwide success.

Monopoly was popular with Darrow's friends, but when he tried to sell the idea to toy manufacturer Parker Brothers (U.S.) the company rejected it, saying that it had "52 fundamental playing errors," including the fact that it was too complicated and took too long to play. But Darrow had faith in the idea, and produced 5,000 sets privately, which sold quickly enough for Parker Brothers to think again; it signed a licensing agreement early in 1935, which soon made Darrow a millionaire.

Parker Brothers sent a sample to British games manufacturer John Waddington Ltd., where Norman Watson (England), manager of the playing cards division, reported that he was "enthralled and captivated" by the game. So much so that he persuaded his father, Victor, who was managing director, to make the company's first transatlantic phone call and secure a licensing deal. The deal signed, Watson Sr. sent his secretary, Marjory Phillips, to walk around London and choose street names for the British version of the game. From that first variation, Monopoly has since been sold in more than 80 countries in 26 languages, and set in cities including Athens, Madrid, Paris, Moscow, and Tokyo.

Left Charles Darrow playing the game that made him a millionaire **Above** Monopoly world championship in Monaco (September 1996)

The patentee: Charles B. Darrow

1889 Born on August 10 in Philadelphia and later works as a plumber and heating engineer

1929–33 After the Wall Street crash, finds employment doing a number of jobs including mending irons, walking dogs, and mowing lawns

1933–34 Develops his version of Monopoly and begins manufacturing his own sets, first for his friends and then commercially. Approaches Parker Brothers (U.S.), which rejects the idea

1935 Signs a licensing deal with Parker Brothers. Files a patent for Monopoly on August 31 (granted December 31). Parker Brothers sends a sample to John Waddington Ltd. (Britain)

1936 Retires at the age of 46 to pursue his interests in ancient cities, cinematography, and collecting exotic orchids. Waddington's begins producing a British version of Monopoly, set in London

1967 Dies on August 28, aged 78, having become the first millionaire games designer

1970 A memorial plaque is erected in Atlantic City, New Jersey, close to the junction of Boardwalk and Park Place (the equivalent of Mayfair and Park Lane in the British version)

Did you know?

The world's most expensive Monopoly set, worth £1.15 million, was created in 1988 by San Francisco jeweller Sidney Mobell (U.S.). It had a 23-carat gold board, solid gold houses and hotels with rubies and sapphires set in the chimneys, and dice with 21 diamonds each for spots.

In the U.S., when the Atlantic City authorities proposed renaming Baltic Avenue and Mediterranean Avenue in 1972, there was such an outcry from Monopoly fans that the names were kept.

Inventors are often accused of trying to reinvent the wheel, but George Ferris Jr. did so with such splendor that his name has survived for more than a century since his big wheel was set up for the Chicago Columbian Exposition in 1893.

FERRIS WHEEL

Ferris wheel has become a quasi-generic term for this archetypal fairground attraction, but the wheel erected by George Washington Gale Ferris, Jr. (U.S.), in Chicago in 1893 was not the first Big Wheel. The first patent for a Big Wheel had been filed 26 years earlier, in 1867, and described two wheels turning side by side. Then, in 1872, Isaac Newton Forrester (U.S.) went one better and invented what he called the Epicycloidal Diversion (patented as a Rotary Swing in 1875), which had four wheels at right angles to each other in a square pattern, all turning vertically about their axis as well as moving horizontally on a rotating base plate.

These and other early Big Wheels were a great attraction, but they paled into insignificance beside the wheel built by Ferris in 1893. Chicago was hosting the Columbian Exposition that year, and the organizers wanted a central attraction to rival the Eiffel Tower, which had dominated the Paris Exposition of 1889. Several ideas were discussed and rejected, and Ferris's wheel almost met the same fate before his audacious plan was accepted: two 132-foot (40-meter) towers supporting an axle that turned a wheel 251 feet (76 meters) in diameter, with 36 enclosed gondolas accommodating 40 seats each—a total of 1,440 seated passengers at a time. To put that in perspective, Forrester's Epicycloidal Diversion had carried a mere 64 passengers.

The original Ferris wheel cost $400,000 to build but took $726,000 in receipts at the Exposition. Despite this phenomenal success, the wheel was not as long-lived as the Eiffel Tower; it was moved to St. Louis, Missouri, for the World's Fair of 1904 and then sold for scrap, fetching a pitiful $1,800. However, Ferris's name lived on, as did the vogue he had started for truly big wheels. More than a century later, the London Eye, currently the world's largest observation wheel, opened in England to celebrate the new millennium—at 116 feet (35 meters), it is nearly twice the diameter of the original Ferris wheel.

Did you know?

In 1900 William E. Sullivan, working for the Eli Bridge Co. (both U.S.), began making Ferris wheels that could be dismantled, which is why big wheels are sometimes referred to as Eli Wheels.

The Columbian Exposition, for which the original Ferris wheel was designed, was organized to celebrate the 400th anniversary of the arrival of Christopher Columbus in the New World.

Like the original Ferris wheel, the London Eye was almost rejected before it was built. It was designed by husband-and-wife team David Marks and Julia Barfield (both Britain) as an entry for a competition organized in 1993 by the *Sunday Times* newspaper and the Architecture Foundation, to find ideas for a monument to mark the new millennium. The prize was not awarded because the judges decided that none of the entries was good enough, but fortunately Marks and Barfield had enough faith in their idea to secure corporate finance and create one of London's most exciting millennium landmarks.

The inventor: George Washington Gale Ferris, Jr.

1859 Born in Galesburg, Illinois, is later educated at high school in Oakland, California

1881 Graduates Rensselaer Polytechnic in Troy, New York, with a degree in engineering and goes on to design several bridges and tunnels

1885 Establishes G.W.G. Ferris & Co. to test the iron and steel leaving foundries for use in bridge and railway construction

1892 Designs the Ferris wheel, which is eventually (and critically) approved by the organizers of the Columbian Exposition

1893 The Ferris wheel opens on May 1, six weeks after the opening of the Exposition

1896 Ferris dies of typhus on November 22, aged 37

1904 The Ferris wheel is moved to St. Louis for the World's Fair and subsequently sold for scrap

Opposite top The original Ferris wheel dominates the Midway Plaisance at Chicago's 1893 Columbian Exposition **Below** Children playing on an improvised Ferris wheel near Mount Carmel in Haifa, Palestine (now Israel) (c. 1900)

Arcade games began with 19th-century penny-in-the-slot machines and have since progressed to the latest high-tech computer games, but the most iconic and long-lasting of all arcade machines is pinball, in all its glorious variations.

Arcade games & pinball time line

1871 Henry Davidson (Britain) invents Chimney Sweep, the first amusement arcade game

1885 William T. Smith (U.S.) invents America's first arcade machine, The Locomotive, a working model of a steam train

1889 Charles Frey (Germany–U.S.) invents the fruit machine

1891 Thomas Edison (U.S.) invents the kinetoscope peephole viewing-machine

1930 David Gottlieb (U.S.)

PINBALL

The first amusement arcade game was a penny-in-the-slot machine called Chimney Sweep, invented in 1871 by merchant seaman Henry Davidson (Britain). Such games became ever more popular, and the concept of the penny arcade was given a huge boost with the invention of the fruit machine by Charles Frey (Germany–U.S.) in 1889 and the kinetoscope by Thomas Edison (U.S.) in 1891. Another 19th-century influence on the invention of pinball was bagatelle, a parlor board game that involved rolling balls into holes to score points. The concepts of arcade machines and bagatelle came together as pinball in 1931, in the forms of a short-lived game called Whiffle and the groundbreaking Baffle Ball, invented by David Gottlieb (U.S.) after he bought the rights to a version of bagatelle called Bingo.

Baffle Ball was a game in which small nails or pins surrounded the scoring pockets—a feature that in 1936 gave rise to the generic name "pinball" for this type of game. Early pinball games were designed to sit on a counter, rather than being self-supporting, and there were none of the trappings associated with modern machines; players even had to tot up their own scores based on which pockets the balls dropped into. Even so, Gottlieb sold more than 50,000 games in 1931, a success rate that inspired one of his distributors, Raymond Maloney (U.S.), to invent his own pinball game, Ballyhoo, and form the Bally Company to market it.

Other innovations quickly followed. Players would often cheat by tipping the machines, so in 1934 Harry Williams invented the antitilt mechanism, originally known as a stool pigeon, which comprised a metal ball that would fall onto a metal ring if the machine was tilted or jolted, completing a circuit and ending the game. Bumpers were invented in 1936, flippers in 1947, and pinball entered the computer age in the 1990s with the introduction of "virtual" pinball played on a computer simulator.

See also: Phonograph/Kinetoscope, page 158

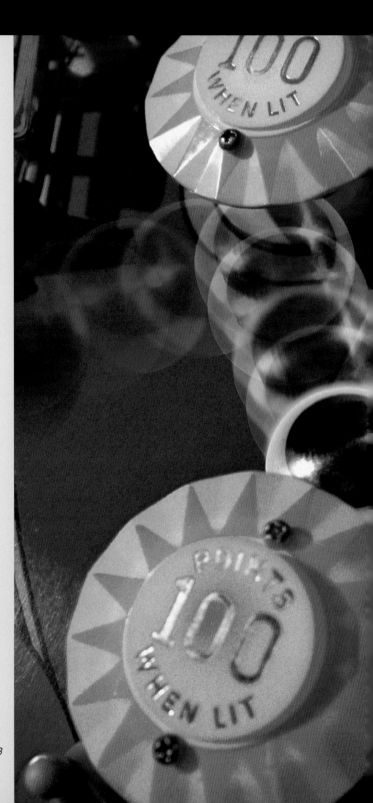

buys the rights to manufacture the amusement Bingo

1931 Gottlieb invents Baffle Ball. A short-lived game called Whiffle, released earlier in 1931, is arguably the first coin-operated pinball game, but Gottlieb's marks the start of the pinball industry

1932 Raymond Maloney (U.S.) invents Ballyhoo and founds the Bally Company. His sales slogan is: "What they'll do through '32...play Ballyhoo"

1934 Battery-operated pinball machines are introduced, with colored lights and backlit glass. Harry Williams invents the antitilt mechanism

1936 Bumpers are invented. The term "pinball" is coined

1942 Pinball is banned in New York City (see Did you know?)

1947 Flippers are invented by D. Gottlieb & Co. and introduced in a pinball game called Humpty Dumpty

1971 Nutting Associates (U.S.) builds *Computer Space*, the first arcade computer video game, written by Nolan Bushnell (U.S.)

1990s Virtual pinball, played on a computer, is introduced

Below Multiple exposure of pinball in action **Bottom** While it is banned in the Big Apple, sailors make the most of playing pinball at the Royal Hawaiian Hotel on Waikiki Beach, Honolulu, Oahu, Hawaii (April 1942)

Did you know?

On January 21, 1942, pinball was banned in New York City after it was classified as gambling because the authorities considered it a game of luck rather than a game of skill. To signal his support for the ban, the then mayor, Fiorello Henry La Guardia (after whom the airport is named), publicly smashed a number of pinball machines. The ban was lifted in 1976, although "free" games, such as replays, are still technically illegal in New York, according to a law that is never enforced.

As well as pinball, the quest for cheap entertainment during the Depression also gave rise to the perennially popular board games of Scrabble, invented as Lexico in 1931 by Alfred Mosher Butts (U.S.), and Monopoly, patented by Charles Darrow (U.S.) in 1935. *See also: Monopoly, page 174*

The term "stool pigeon," usually used to describe a police informer but also used at one time for pinball antitilt mechanisms, derives from a hunting decoy that in its original form was literally that—a pigeon stuck to a stool.

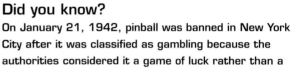

Roller skates were invented by Joseph Merlin for an 18th-century masquerade performed in London, but they failed to become popular. Surprisingly, in-line skates were invented as early as 1823, almost 40 years before four-wheeled skates.

ROLLER SKATES

Roller skates made their first public appearance in 1760 at Mrs. Cornelly's masquerade at Carlisle House in Soho Square, London, England. They were the invention of instrumentmaker Joseph Merlin (Belgium), who made a spectacular entrance, playing the violin and wearing skates that ran on wooden spools. It is perhaps not surprising that his invention failed to start a new trend, because he swept gracefully across the ballroom into a huge mirror, smashing the mirror, his violin, and himself. Merlin survived, but his injuries were not a great advertisement for roller-skating.

Sixty-three years later, on April 22, 1823, Robert John Tyers (England) was granted a patent for "apparatus to be attached to boots...for the purpose of travelling or pleasure." These roller skates, which Tyers called Volitos, were a conscious imitation of an earlier invention, each skate having five wheels in a single line to imitate the blades of ice skates, and were first demonstrated by Tyers on a tennis court in Windmill Street, Soho, London. Tyers had a slightly greater number of imitators than Merlin, and a similar type of skate was used in 1849 for the first production of Jakob Meyerbeer's opera *The Prophet*, which called for ice-skating in the third act: In-line skates were used as a cheaper alternative to creating an ice rink onstage.

But roller-skating really took off only after James Leonard Plimpton (U.S.) invented four-wheeled roller skates in 1862 (patent granted January 6, 1863, U.S.). The wheels were made of wood with a special mechanism for turning: "The rollers are runners made to turn or cramp like the wheels of a wagon by the rocking or canting of the stock or footstand to facilitate the turning of a skate." Plimpton, who had already patented a fastening for ice skates, called his new invention "parlor skates" because he intended them to be used indoors or on a rink. In 1866 he opened the world's first roller rink, in Newport, Rhode Island, and started a national craze for roller-skating.

Above Members of the U.S. women's roller-skating team training for the Harringay Roller Derby (May 1953) Below In-line skating on a ramp at Bondi Beach, Sydney, Australia (March 1995)

Roller-skating time line

1760 Joseph Merlin (Belgium) invents roller skates, but his demonstration at a masquerade in London ends disastrously

1823 Robert John Tyers (England) is granted a patent for in-line roller skates

1840 A beer hall in Berlin, Germany, puts its barmaids on roller skates

1849 Roller skates are used in a production of Jakob Meyerbeer's opera *The Prophet* in Paris, France

1863 Plimpton is granted a U.S. patent for the first four-wheeled roller skates (U.K. patent granted 1865)

1866 Plimpton is granted a patent for an improved turning mechanism, and opens the first roller rink, at the Atlantic House Hotel in Newport, Rhode Island

1869 A.J. Gibson (U.S.) is granted a patent for the invention of circular plates that enable the wheels to be steered to facilitate changing

direction, rather than merely canting as in Plimpton's 1866 invention

1875 The Belgravia Skating Rink in London opens as England's first roller rink

1878 The game of Rink Polo (later Roller Hockey) is invented

1880s Micajah Henley (U.S.) patents several inventions including brake pads, adjustable bindings, and, with his nephew Robert Henley, ball bearings to improve turning

1936 Britain wins the inaugural Roller Hockey World Championship

Technically speaking, Jet Ski is not a description of an invention—it is a trademark of the Kawasaki company. The generic term for this type of craft is personal watercraft (PWC), a concept invented by Clayton Jacobson II in 1965.

JET SKI

In the early 1960s Clayton Jacobson II (U.S.) was an overworked banker who rode dirt bikes as a hobby. Some 30 years later, in an interview for *Personal Watercraft Illustrated* (a magazine that would not have existed were it not for Jacobson's invention), he said: "It was a form of stress relief for me. However, as you know, when you crash a dirt bike, the ground isn't very forgiving. That's why and how I came up with the idea for a personal watercraft. I was looking for a softer landing, and the water offered exactly what I was looking for. Sort of a motorcycle for the water."

Like many great inventions, it was a simple idea that was not so simple to realize. Jacobson built his first working prototype in 1965—a stand-up craft with a one-piece aluminium hull and a fixed, upright handle—followed in 1966 by a second, fibreglass prototype. He then agreed to a licensing deal with Bombardier Corp. (U.S.), the makers of Ski-Doo snowmobiles, and worked with Bombardier to develop the Sea-Doo, the first production personal watercraft. The original Sea-Doo was not a great success and was scrapped in 1970 (a later version has since proved very popular), but Jacobson had faith in his invention and wasted no time in resurrecting the idea. In 1971 he signed a new agreement with Kawasaki (Japan), which resulted in the commercial launch of the Jet Ski watercraft in 1973.

A global leisure and sport industry has grown from Kawasaki's two 1973 models, one of which had a flat hull for stability, the other a deeper V-hull for competitive riders. Kawasaki engineer Fred Tunstall (U.S.) later told *Boats.com*: "I think most of us who worked on the project knew we had something with this product. Once you spent some time with the machines and got used to riding them, they were a lot of fun and tended to make enthusiasts even of the people working on them....But knowing the sport would get this big, that it would all lead to this? I don't think anyone had an idea that would happen."

Did you know?

The name Jet Ski now seems so appropriate for this invention that it is often used (erroneously) to describe any type of personal watercraft. However, at first, even Clayton Jacobson and Kawasaki didn't know what to call it—some early manuals referred to the craft as a Power Ski and some as a Water Jet.

Jacobson described the personal watercraft as "a motorcycle for the water," but the sport has since been described as waterskiing without a boat.

The inventor: Clayton Jacobson II

19?? Little is known about Jacobson's early life

1960s A motocross and dirt-bike enthusiast, he conceives of a "motorcyle for the water"

1965 Builds the first prototype personal watercraft

1966 Builds a second prototype. Signs a licensing agreement with Bombardier Corp. (U.S.), which results in the Sea-Doo

1971 Signs a licensing agreement with manufacturer Kawasaki (Japan)

1973 Kawasaki launches two models of the Jet Ski watercraft, both powered by a 398cc Kawasaki snowmobile engine

Above A Jet Ski is the aquatic equivalent of a speeding motorcycle, but crashing is much less dangerous
Below Tim Tynon demonstrates a daredevil wave-jumping maneuver at Imperial Beach, California (1990)

Life without this invention would be unthinkable for most DIY enthusiasts and many professionals, but, as with so many great inventions, a number of companies rejected the idea when Ron Hickman first tried to license his workbench in 1967.

WORKBENCH

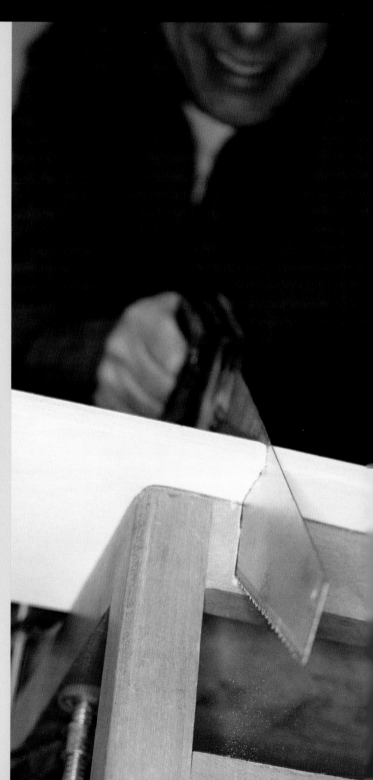

Despite the fact that he had no formal engineering or design qualifications, Ron Hickman (South Africa) was a very successful design engineer for Lotus Cars (England), being principal designer of the Elan (1962), the Elan Plus 2 (1966–67), and the Europa (1965). The Elan was the first production car to feature pop-up headlights and Hickman's invention of body-contoured plastic bumper panels. After nine years with Lotus, Hickman left the company to pursue his own ideas, one of which was the workbench that would later achieve world fame as the Black&Decker Workmate workbench.

One weekend in 1961, while doing some DIY, Hickman had used a Swedish whitewood chair as a sawhorse on which to cut a piece of plywood. When he realized that he had nicked the chair while sawing the plywood it seemed like an irritating mistake, but in fact it was to make him a multimillionaire. He conceived the idea of a portable, folding workbench and developed this idea over the next six years. He puts his success down to the fact that he was not a trained designer, so he approached the project with no preconceptions about what a workbench should look like. He had two narrow pieces of wood handy for the prototype rather than a single wide one, and this led to the defining feature of the Workmate: the two separate beams that act both as a vice and a worktop.

In 1967 he tried to sell the manufacturing rights to his invention but was rejected by eight companies, including Black&Decker and Stanley Tools. Undeterred, he filed patents for his invention in 1968, established Mate-Tools Ltd., set up a factory in an old brewery, and began manufacturing and selling his workbench. Sales rose to more than 12,000 four years later, prompting Black&Decker executives to realize they had made a mistake. In 1972 they agreed to a licensing deal that saw Hickman retire to Jersey as a tax exile and Black&Decker still selling the workbench more than 30 years later under the trademark Black&Decker Workmate workbench.

Did you know?

A later invention by Ron Hickman was the Footprint Potty, a child's potty with an extended front section incorporating footprints for the child's feet. Sometimes wet skin sticks to plastic potties, causing the potty to tip up and spill—Hickman's idea, which worked but did not sell, was that as the child stood up, its weight on the front section would keep the potty from tipping. Unfortunately, the marketing did not explain the purpose of the invention, so the potty was not a commercial success.

As a private vehicle design project in the 1950s, Hickman invented a novel flip-top roof/door concept. He later found out that his friend, designer Tom Karen (Czechoslovakia–Britain, b. Austria), had invented a similar concept—the fiberglass Bond Bug.

Opposite top An early version of Ron Hickman's versatile workbench, the forerunner of the world-famous Black&Decker Workmate workbench
Below It was a mistake like this that inspired Ron Hickman to create the folding workbench

The inventor: Ron Hickman

1932 Born Ronald Price Hickman, the second of five children, on October 21 in Greytown, Natal, South Africa

1937–48 Educated to matriculation level at four successive local schools

1949–54 Works for the South African Department of Justice (Magistrate's Division)

1955 Moves to England and works as styling modeler for Ford, for whom he invents and designs the world's first body-shaped plastic bumper panels

1958–67 Works for Lotus Cars (England), from 1960 as road cars chief of design & development

1961–67 Invents what he calls the Mini-Bench after damaging a chair while doing some DIY. Tries to license the idea but is rejected by eight companies

1968 Files a patent on March 4 for his workbench and subsequently files 15 further patents for improvements. Establishes Mate-Tools Ltd. to manufacture and sell the benches under the trademark Workmate

1972 Black&Decker, which had originally rejected his idea, agrees to a licensing deal to manufacture and market the bench under the trademark Workmate

1973 Hickman wins a Design Council award for the dual height Mark II Workmate, the first time the award has been presented for a tool

1977 Moves to Jersey, where he invents or backs more than 50 products, including the Footprint Potty, the Peeping Tom inspection mirror (an extendable, hinged inspection mirror with lights, for dark or awkward places), and the Hi-Mate folding ladder. (None of his inventions has been as successful as the Workmate, but Hickman says: "Once an inventor, always an inventor")

1974–82 Successfully sues 20 patent infringers

1980–83 Designs his own family house, Villa Devereux, which later features on a Jersey postage stamp

1994 Awarded the OBE (Officer of the Order of the British Empire) for Services to Industrial Innovation

2000 Features on a Jersey millennium phonecard

Did you know?

Stanley Tools rejected Ron Hickman's invention because the company said that potential sales "would be measured in dozens rather than hundreds." By 2003 Black&Decker had sold more than 65 million Workmates worldwide.

The development of chainsaws—from electric machines for use in sawmills, to gas-powered saws that could be used in the forest, to a one-man chainsaw—was a game of leapfrog between two companies, Dolmar and Stihl.

CHAINSAW

The modern definition of a chainsaw is "a portable power-driven saw with teeth linked together in a continuous chain," but the first chainsaw, invented in 1926 by Andreas Stihl (Germany), was far from portable. It was an electrically powered machine that improved the efficiency of sawmills and timber yards but was very different from the chainsaw we know today. The following year Emil Lerp (Germany) took a significant step forward with the invention of a gas-driven chainsaw, which he demonstrated on the wooded slopes of Mount Dolmar in the Thuringia region of Germany. Thuringia is a densely forested region, so it is apt that it is recalled in the name of one of the largest chainsaw manufacturers in the world, Dolmar GmbH (now based in Hamburg), which was founded by Lerp and named after the mountain where he first demonstrated his invention.

Stihl followed Lerp's lead in 1929 with his own gas-driven chainsaw, whose name translates as the Stihl Tree-felling Machine, and then set his mind to developing a gas-driven saw that could be operated by one man. During the next 20 years gas engines became small enough to make that a possibility, and in 1950 Stihl produced the Stihl BL, a one-man chainsaw that was the first example of the chainsaw as we know it today.

In February 1997 using a chainsaw was made easier and safer by forestry worker Charles Brathwaite (Barbados, resident in Wales) and retired blacksmith Idris Jones (Wales). Together, they produced the Logmaster Clamp, an invention first conceived by Brathwaite in 1987 as an ingenious, easy-to-use clamp mounted on an adjustable tripod that made a cumbersome workhorse unnecessary. Apart from the fact that it is lightweight and collapsible, the greatest advantage of Brathwaite's clamp over a traditional workhorse is that the log or beam is sufficiently securely held with one clamp only, making the entire sawing process much safer.

Above Charles Brathwaite demonstrates his invention, the Logmaster Clamp **Opposite** An ice fisher cuts a hole into frozen Lake Winnebago, Wisconsin, with a chainsaw (March 1988)

Logmaster Clamp inventor: Charles Brathwaite

1967 Born on June 10 in Bridgetown, Barbados, and later educated at the Lodge School, Messiah St., St. Johns, Barbados

1983 Moves to Britain

1987 Conceives the Logmaster Clamp and makes his first drawings

1997 Works as a self-employed landscape gardener. Produces the first Logmaster Clamp with Idris Jones (Wales)

2000 Files a patent for the clamp on October 4 (patent pending)

2001 Wins the Welsh Development Agency's Inventor of the Year competition

2002 Approaches inventors' agency Inventorlink to publicize and secure financial backing for the Logmaster Clamp

The crossword puzzle was invented by Arthur Wynne in 1913 as a quick and simple word game, but soon developed into a devilish brainteaser with the invention, in 1925, of the cryptic crossword by Edward Powys Mathers, aka "Torquemada."

Crossword puzzle time line

1913 Arthur Wynne (England) invents the Word-Cross Puzzle, first published on December 21 in *New York World*

1914 Name changed to crossword puzzle

1921 Margaret Petherbridge

(U.S.) introduces single-number clues

1922 News of crosswords reaches Britain, where the February edition of *Pearson's Magazine* describes "a new type of puzzle in the shape of a word square....These new

CROSSWORD PUZZLE

Like most inventions, the crossword did not simply occur to the inventor from thin air, but was inspired by an earlier idea. Arthur Wynne (England) was working in what was known as the Tricks & Jokes Department of the *New York World* weekend supplement, his job being to devise new puzzles and games to entertain readers. He remembered playing a Victorian parlor game—known as Magic Square or Double Acrostic—with his grandfather, and he used this game as the basis for what he called a Word-Cross Puzzle. The grid style and the clue-numbering system would later change, but Wynne had invented the world's first crossword puzzle, and it was published in *New York World* on Sunday December 21, 1913.

By January 1914 the name of Wynne's invention had been reversed to crossword, and in 1921 Margaret Petherbridge (U.S.), *New York World*'s new crossword editor, changed the way the clues were numbered. Wynne had placed a number at the beginning and end of each solution, and the clues were numbered accordingly (2-3, 13-21, etc.), but Petherbridge replaced this system with the single numbers common today. By 1924 the clues were being listed separately as Horizontals and Verticals, thought to be the innovation of C.W. Shepherd (England), compiler of Britain's first newspaper crossword, which was published in that format on November 2, 1924, by the *Sunday Express*.

In 1925 the invention of the cryptic clue took crosswords to an entirely new level. The first cryptic crossword was compiled by Edward Powys Mathers (England, aka "Torquemada" after the Spanish inquisitor) and appeared in the *Saturday Westminster*. Cryptic crosswords immediately became popular in Britain but not in America, where one crtic complained about the "puns, anagrams, rare literary allusions, out-of-the-way place names, all-but-forgotten cricket and soccer terminology, and downright unsporting tricks"—an accolade that would no doubt have delighted Torquemada.

word squares are having a tremendous vogue in America just now"

1924 Simon & Schuster (U.S.) publishes, under the imprint Plaza Publishing, the first book of crossword puzzles, which becomes a bestseller. On November 2 the *Sunday Express* publishes Britain's first newspaper crossword

1925 A Broadway revue (U.S.) features a sketch in a fictitious "Crossword Puzzle Sanitarium." On July 1 *Punch* magazine (Britain) reports that the crossword has been "the dominant feature of our lives in these last six months. Its appeal to the British bosom has been universal. It touched all classes, because it made demands upon a modicum of intelligence common to them all." Torquemada (aka Edward Powys Mathers [England]) demands more than a modicum of intelligence by inventing the cryptic crossword, first published by the *Saturday Westminster*

1930 On February 1 *The Times* outrages many of its readers by publishing the first *Times* crossword, compiled by Adrian Bell (England). One correspondent writes: "I hate to see a great newspaper pandering to the modern craze for passing the time in all sorts of stupid ways"

Below Some clues can be real head-scratchers **Bottom right** Fledgling publishers Lincoln M. Schuster and Richard Simon with their 1924 book of crossword puzzles

Did you know?

During the crossword craze of the 1920s the British Optical Association raised concerns that puzzling over crosswords would cause eyestrain and headaches. On the other hand, the Chicago Department of Health announced that crosswords provided a mental stimulus that promoted health and happiness.

The late Princess Margaret (England) won prizes for completing the crossword in *Country Life* magazine.

The editor of *The Times* crossword was cross-examined by police in October 1966 after a reader reported the fact that the words "gaol" and "artillery" appeared in the crossword two days before spy George Blake escaped from Wormwood Scrubs jail into the adjacent Artillery Road.

C.W. Shepherd became Britain's first newspaper crossword compiler after selling an American crossword to the *Sunday Express*. The day before publication the editor noticed that one of the solutions was the word "honor," spelled the American way. Fitting in the *u* was no simple matter, and Shepherd ended up compiling a new crossword altogether— without the word "honour."

AND ALSO...

Tiddlywinks

It is hard to imagine that a game as simple as Tiddlywinks was actually invented, rather than simply evolving, or that anyone took the trouble to patent it, but Joseph Assheton Fincher (England) did so on November 8, 1888, with what patent curator Stephen van Dulken describes as "one of the shortest and simplest patents ever." The following year, Fincher even registered Tiddlywinks as a trademark, but the existence of the patent and the trademark did not stop countless variations of the game being patented and still more being sold. This did not dampen Fincher's inventive fervor, however, and he was issued two further patents, one in 1890 for cufflinks and another in 1897 for candlesticks.

Apparatus for walking on water

On January 5, 1915, Marout Yegwartian (England) was granted a patent for an invention that resembled a cross between two boats and a pair of oversize shoes, which was intended to enable people to walk on water. His patent stated: "This invention refers to and consists of new or improved water-treading appliances for enabling persons to cross or navigate a stream, river, or the like on foot, the invention being particularly useful for campaigning and pioneering purposes, although also useful for sporting and pleasure purposes." It all sounds highly improbable, but in fact Leonardo da Vinci had invented a very similar device more than 400 years earlier. Da Vinci's "flotation shoes" were successfully tested in the 19th century by aviation pioneer Lawrence Hargrave (England–Australia), and in 1988 Remy Bricka (France) succeeded in using similar apparatus to "walk" across the Atlantic. He left the Canary Islands on April 2, 1988, and arrived in Trinidad on May 31, having made his 3,380-mile (5,634-kilometer) crossing on two 14-foot (4.2-meter) floating polyester skis.

Pat-on-the-back apparatus

On May 31, 1985, Ralph R. Piro (U.S.) filed a patent for a mechanical hand that could be fixed to a person's shoulder in order to pat them on the back (patent granted 1986). Perhaps the reason Piro thought this was a practical invention is that he was traumatized as a child by Thomas V. Zelenka's baby-patting machine (*see page 44*). Piro's patent specifies that his invention "can be utilized to promote feelings of well being necessary to a positive mental attitude. Such an arrangement may provide the needed psychological lift to allow a person to overcome some of the 'valleys' of emotional life in a highly technicalized society that often postpones the level of immediate personal approval desirable for continued accomplishment."

Postcard

The postcard was invented and copyrighted in 1861 by John P. Charlton (U.S.), who sold the rights to stationer Hyman L. Lipman (U.S.). In 1868 Emmanuel Herrmann (Austria) went a step further by inventing the prepaid postcard, an idea published in *Neue Frei Presse* that January and realized the following year when the first *Correspondenz-Karte* was issued on October 1, 1869, by the Austrian post office. The *picture* postcard was invented later still, in 1872, when J.H. Locher (Germany) published three cards, the first showing six small views of Zürich engraved by Franz Rorich (Germany) and the next two each showing three views.

Rubik's Cube

Rubik's Cube was invented in 1974 by Erno Rubik (Hungary), a professor of interior design in Budapest, and patented by him in 1975. It was an intriguing and hugely successful puzzle, but, after signing a licensing deal to manufacture the cube in Western Europe and the U.S., Ideal Toy Co. discovered that it was not the first such cube. Larry Nichols (U.S.) successfully sued Ideal

Opposite top Contestant playing at the International Tiddlywinks Championships, San Francisco (August 1962)
Above Lighted handbags may make blind digging a thing of the past

Toy for patent infringement, citing a patent he had filed in 1972 for a similar, magnetic cube.

Slinky

In principle the Slinky is nothing more than a spring, but in practice it took two years to find the right tension, diameter, and gauge of wire to make this toy slither downstairs, or across the floor. It was conceived in 1943 and perfected in 1945 by U.S. Navy engineer Richard James (U.S.), who had seen an engine spring slide off a table on a rolling ship and was inspired to think that a similar spring could make a good toy. His wife, Betty, coined the name Slinky (which was registered as a trademark in Britain in 1946 and the U.S. in 1947), and together they demonstrated the new invention at Gimbel's department store (U.S.) in November 1945, where they sold all 400 samples for one dollar each.

Handbag light

In 2003 handbag manufacturers Philipp and Axel Bree (Germany) invented a handbag light known as the Bree Light, an ingenious device that acts like the light in a fridge, coming on when the bag is opened. Axel said: "We had been toying with the idea of illuminating the dark insides of handbags for quite some time, but lacked an elegant solution which not only met our high demands with regard to style and function, but was also technically feasible." Journalist Stefanie Marsh reported in *The Times* that the invention "will not only illuminate [the bag's] contents, but may even offer scientists clues as to how a woman can fit the entire contents of her kitchen, bathroom and garage into a space no bigger than a cushion."

Chapter Six

AT THE DOCTOR'S

Pharmacological expertise and an inventive streak came together in 1964 when James Black invented the first practical beta-blocker. Working to the same principle, he followed this in 1972 with a vital new histamine blocker.

BETA-BLOCKER

Certain functions of the body are controlled by "messenger" molecules, which are released by some nerves and organs to provoke a response in other nerves and organs that have "receptor" cells for that messenger. One thing that confused scientists was that some messengers provoke diametrically opposed responses: For example, adrenaline causes some muscles to relax and others to contract.

Pharmacologist James Black (Scotland) unraveled this mystery while researching a treatment for angina. He had noticed that although existing adrenaline blockers succeeded in relaxing blood vessels, they had no effect on reducing heart rate—this made no sense, because it was well known that adrenaline increased heart rate, and therefore, blocking adrenaline should reduce it. Researching further, he discovered that Raymond Ahlqvist (U.S.) had theorized in 1948 that there were two types of adrenaline receptor: alpha and beta. Black deduced that the existing adrenaline blockers must be alpha-blockers, and that the heart must rely on beta-receptors, so he set about developing a beta-blocker that could treat angina by nullifying the effect of adrenaline on the heart and thereby slowing it down.

The first beta-blocker was developed by a rival research team in 1958 but had such extreme side effects that it could not be used clinically. Black's team produced one safe beta-blocker, pronethalol, in 1962, and improved it to produce propranolol, the world's first clinically practical beta-blocker, in 1964. Continuing to research the implications of inhibiting receptor cells, in 1972 Black identified another important receptor and subsequently developed an appropriate blocker. The primary cause of stomach ulcers is overactive acid secretion, which is stimulated by the messenger histamine. Black discovered that, as with adrenaline, there were two types of histamine receptor, H1 and H2; by developing an H2 blocker, he provided medical science with an effective treatment for stomach ulcers.

The inventor: Sir James Black

1924 Born James Whyte Black in Uddingston, Scotland

1946 Graduates in medicine from St. Andrews University, Scotland

1946–47 Assistant lecturer in physiology at St. Andrews University

1947–50 Lectures at the University of Malaya

1950–58 Becomes senior lecturer at the University of Glasgow, Scotland, where he establishes a new department of physiology

1958–64 Works for ICI Pharmaceuticals (England)

1962 Produces the first safe beta-blocker—pronethalol

1964 Produces the first clinically practicable beta-blocker—propranolol

1965 Works for Smith, Kline & French Laboratories (England)

1972 Identifies the H2 group of histamine receptors

1973–77 Professor of pharmacology at University College, London, England

1976 Elected Fellow of The Royal Society

1977–84 Becomes director of therapeutic research at the Wellcome Research Laboratories, where he develops the H2-blockers (histamine-blockers) cimetidine and burimamide

1981 Receives a knighthood

1985 Appointed a professor at King's College Medical School, London

1988 Shares the Nobel Prize for Physiology or Medicine

1992 Becomes Chancellor of Dundee University, Scotland

Did you know?

James Black shared the 1988 Nobel Prize for Physiology or Medicine with George Hitchings and Gertrude Elion (both U.S.). Hitchings and Elion worked together to develop compounds to inhibit DNA synthesis for use in cancer treatment, and their research led to treatments for leukemia, malaria, gout, and kidney stones, and to the development of immune repressors useful in transplant surgery. They also produced a compound to treat the herpes virus, which led to the development of the drug zidovudine (AZT), which prolongs life and alleviates symptoms in AIDS sufferers.

Opposite top An office worker suffering from angina, his office showing three of the main causes: overwork, stress, and smoking
Below A polarized light micrograph of the beta-blocker propranolol, developed by James Black and his team

X-ray photography was such a huge step forward for medical diagnosis that Wilhelm Konrad von Roentgen, who invented the technique in 1895, was awarded the first Nobel Prize for Physics when the prizes were inaugurated in 1901.

X-RAY PHOTOGRAPHY

X-rays were not an invention, they were a discovery. This form of radiation already existed in nature before it was discovered by Wilhelm Konrad von Roentgen (Germany) in 1895—what Roentgen invented was X-ray photography, aka radiography, a practical application for his discovery. Late in 1895, Roentgen, then Professor of Physics at the University of Würzburg, Germany, was investigating the known phenomenon of cathode rays by passing electricity through various gases. On the evening of November 8, he noticed that a piece of paper coated with a fluorescent chemical glowed even when a piece of cardboard physically separated the paper from the source of the rays. The paper continued to glow when Roentgen took it into the next room, and he realized he had discovered a new form of radiation.

Roentgen called his discovery X-rays because at that stage he did not know what they were, X being the scientific symbol for an unknown factor. He continued his experiments for another month, and discovered that not only would the rays pass through solids, but a photographic plate would capture an image of whatever lay between the source of the rays and the plate. On December 22 he took the now-famous X-ray of his wife's hand, showing her wedding ring and the bones of her fingers, and on December 28 he presented his findings to the Würzburg Physico-Medical Society in a paper entitled "On a New Kind of Rays."

News of Roentgen's discovery soon spread around the world, revolutionizing medical diagnosis. Before this, the best means of working out what was happening inside the body had been manipulation ("tell me where it hurts"), auscultation (listening to the chest) and percussion (tapping the body to ascertain the health of the organs). Roentgen's discovery, and the invention of radiography, was such a profound step forward that it was little surprise when in 1901 he was awarded the inaugural Nobel Prize for Physics.

See also: CAT scanner, page 222

Above Portrait of Wilhelm Konrad von Roentgen (c.1895)
Opposite X-ray image of the lumbar spine and pelvis

The inventor: Wilhelm Konrad von Roentgen

1845 Born in Lennep, Prussia (now Remscheid, Germany) on March 27, he later studies mechanical engineering in Zürich, Switzerland

1879 Appointed Professor of Physics at the University of Giessen, Germany

1888 Appointed Professor of Physics at the University of Würzburg, Germany

1895 Discovers X-rays (aka Roentgen Rays) and invents technique of X-ray photography

1896 Awarded the Rumford Medal by The Royal Society for his work on X-rays

1899 Appointed Professor of Physics at the University of Munich, Germany

1901 Awarded the first Nobel Prize for Physics for his work on X-rays

1923 Dies on February 10, in Munich, aged 77

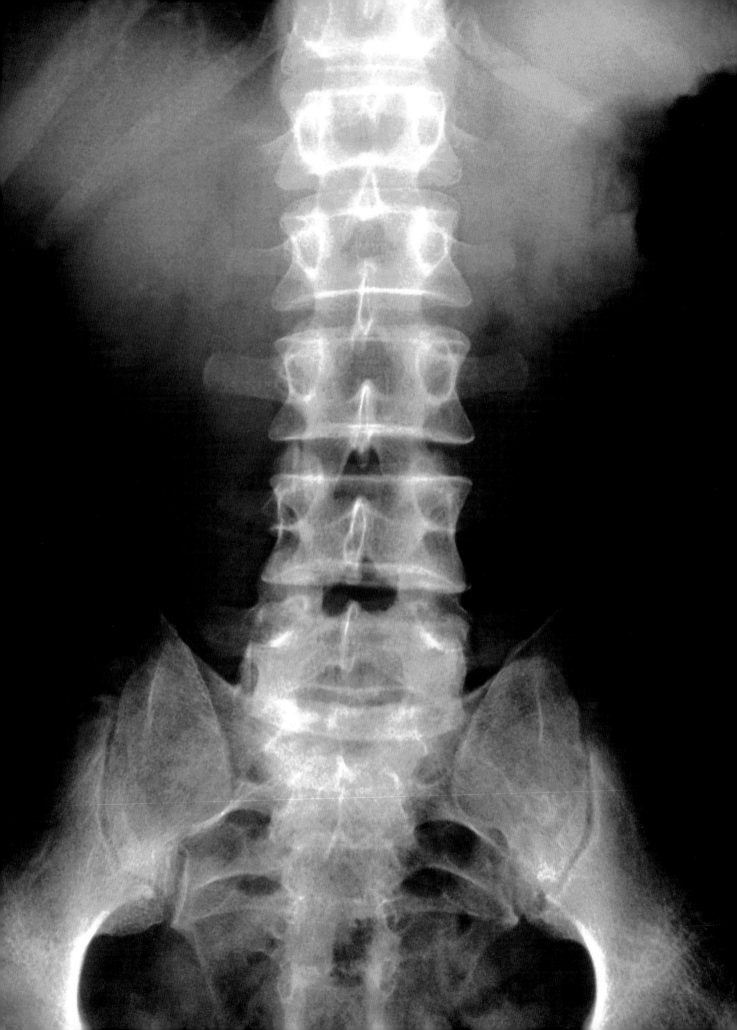

Oral contraceptives, more popularly known as the Pill, were first made available to the public in 1960. The Pill set the tone for the "swinging sixties," but the research that led to its invention by Gregory Pincus dates back to 1919.

THE PILL

The contraceptive pill was developed by Gregory Pincus and John Rock (both U.S.) during the 1950s, but the synthetic hormones that form its active ingredients were the result of much earlier research. The first person to suggest the idea of a contraceptive pill was Ludwig Haberlandt (Austria), who began his research at the University of Innsbruck in 1919: He discovered that hormones could be used to prevent pregnancy, and in 1927 he declared: "My aim: fewer but fully desired children!"

Building on Haberlandt's research, scientists discovered that the active hormone was progesterone, but this was expensive to isolate and largely ineffective when taken orally. Then, in 1939, Russell Marker (U.S.) discovered that progesterone could be synthesized from a chemical contained in a particular type of Mexican yam, and formed a company called Syntex to further his research. A decade later, Carl Djerassi (Austria–U.S.) took up an invitation to work for Syntex, and in October 1951 Djerassi and his colleagues Luis Miramontes and George Rosenkranz (both Mexico) succeeded in creating the first synthetic progesterone.

Meanwhile, Frank Colton (Poland–U.S.) had also been building on Marker's research, and in 1953 he filed two patents relating to another method of synthesizing hormones. Colton and Djerassi were both using their synthetic hormones to treat menstrual problems, and it seems that no one thought of using them as contraceptives until Margaret Sanger (U.S.) of the planned parenthood movement initiated research into an oral contraceptive. As a result, Pincus and Rock developed a pill (using Colton's product) that they field-tested from 1954 onward; it became commercially available in the U.S. in 1960, and in Britain the following year. Inevitably, the Pill has caused ethical controversy, not least because it enables women to live out the words of Carl Djerassi, who said: "Sex should be done for pleasure; reproduction for reproduction."

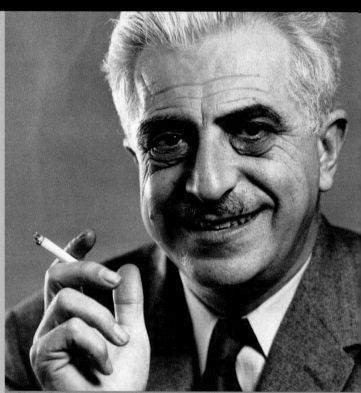

Above Dr. Gregory Pincus (1903-1967), "Father of 'The Pill,'" in 1961 **Opposite** Danny DeVito, Arnold Schwarzenegger, and Emma Thompson in Ivan Reitman's *Junior* (1994)

The inventor: Gregory Pincus

1903 Born Gregory Goodwin Pincus in Woodbine, New Jersey

1924 Graduates from Cornell University

1924–27 Studies as a postgraduate at Harvard University

1927–30 Works at Cambridge University, England, and Berlin University, Germany

1930–38 Member of the biology faculty at Harvard University

1938–45 Member of the Experimental Zoology Department at Clark University

1944 Establishes the Worcester Foundation for Experimental Research, U.S.

1951 Begins developing a contraceptive pill with John Rock (U.S.)

1954 Begins field trials of the Pill

1960 The Pill becomes commercially available

1967 Pincus dies, aged 64

The story of Lorenzo's Oil is the extraordinary tale of a mother and father, Michaela and Augusto Odone, who undertook their own biochemical research in order to invent a treatment for the fatal disease suffered by their son Lorenzo.

LORENZO'S OIL

In 1983 Lorenzo Odone (U.S.) was diagnosed with adrenoleukodystrophy (ALD), a rare genetic disorder that causes the breakdown of a fatty tissue known as myelin (the protective sheath that surrounds the body's nerve fibers), leading to impaired or lost conduction of nerve impulses between the brain and other parts of the body. Doctors told Lorenzo's parents, Michaela and Augusto, that the disease was always fatal and that there was no cure or treatment—Lorenzo would be dead within two years.

Like most parents, the Odones were not prepared to accept this bleak death sentence. *Unlike* most parents, they decided to begin their own biochemical research into the disease, despite the fact that neither of them had any scientific training. They eventually invented a combination of two fats extracted from olive oil and rapeseed oil (glyceryl trioleate and glyceryl trierucate) that stopped the disease in its tracks by preventing the body from producing certain fatty acids, the buildup of which leads to the breakdown of myelin. It was not a cure for ALD, and did not repair already damaged myelin, but it did arrest the progress of the disease. They named it after their son, and Lorenzo's Oil has since become the standard medical treatment for ALD.

Thanks to his parents, Lorenzo defied the doctors' predictions, and celebrated his 25th birthday on May 29, 2003. Sadly, ALD has ravaged his body, but his mind is still active; he communicates by blinking his eyelids to say "no," and wiggling his fingers to say "yes." There is still hope that Lorenzo might regain his speech and full mobility if the Myelin Project, set up by the Odones in 1989, is successful in finding a method that will actually repair myelin, which would be an even greater achievement than the invention of Lorenzo's Oil. The medical profession is still trying to prove scientifically whether or not Lorenzo's Oil works, but for many people the fact that Lorenzo has survived 20 years, and that other sufferers, when treated early, have remained symptom free, is proof enough.

Above "Some people make their own miracles": poster for the film *Lorenzo's Oil* (1992) **Opposite** Augusto Odone, who co-developed Lorenzo's Oil to save his son's life

The inventors: Michaela & Augusto Odone

1933 Augusto is born on March 6 in Rome, Italy

1939 Michaela is born Michaela Murphy on January 10 in Yonkers, New York

1955 Augusto graduates from the University of Kansas with a degree in economics

1960 Michaela graduates cum laude from Dunbarton College of the Holy Cross, Washington, D.C. Michaela is granted a French government scholarship and a Fulbright scholarship to study and teach in France

1966 Michaela and Augusto meet in Milan, Italy. Michaela is named in list of Outstanding Young Women of America

1978 Lorenzo is born on May 29

1983 Lorenzo contracts ALD at the age of five

1985 Michaela and Augusta develop Lorenzo's Oil

1989 A patent for Lorenzo's Oil is filed on September 25 (granted July 19, 1994) by Croda Universal Ltd. (Britain), a chemical company hired to develop one of the components of the oil. After several years of litigation, a judge orders the U.S. Patent and Trademark Office to add Augusto Odone as a co-inventor, which is done on November 3, 1998

1991 Augusto is awarded an honorary Ph.D. for his research into a treatment for ALD

2000 Michaela dies of lung cancer on June 10

2003 Lorenzo celebrates his 25th birthday

The first patent for a medicine was granted in 1698 to botanist and physician Nehemiah Grew for inventing a method of preparing a "purgeing water," which he called Epsom salts after the spa town from whose spring the salts derived.

EPSOM SALTS

Epsom salts are not an invention, being the natural residue of magnesium sulfate left after the evaporation of water from a mineral spring close to the one-time spa town of Epsom in Surrey, England. Because it is a naturally occurring substance, magnesium sulfate is not patentable, but in 1698 Nehemiah Grew (England) was granted the world's first medicinal patent for inventing a method of preparing these "salts" on a commercial scale. Although people were more used to bathing in the Epsom spring waters as a curative for skin conditions, Grew recommended that his preparation should be drunk as a purgative—or, in modern parlance, a laxative.

In Grew's time Epsom was a spa town renowned for its medicinal springs and, as a physician, Grew was naturally interested in what gave the water its medicinal properties. As early as 1695, he described "the bitter cathartic salt of Epsom water," and two years later he published *A Treatise of the Nature and Use of the Bitter Purging Salt Contained in Epsom and Other Such Waters.* He discovered that the key salt was magnesium sulfate, which helped with skin conditions, and that it also had laxative properties—Epsom salts work by preventing the bowels from absorbing water, the overabsorption of which causes constipation. In 1698 Grew was granted a patent for: "The way of making the salt of the purgeing waters perfectly fine in large quantities and very cheape, so as to be commonly prescribed and taken as a general medicine in this our kingdom."

Grew is far better known as a botanist than as a physician, and the "invention" of Epsom salts is rarely mentioned in accounts of his life. His first publication was *An Anatomy of Vegetables Begun,* and he went on to make crucial discoveries in the field of botany, including the fact that plants reproduce sexually and that the stamen is the male organ of the plant. His *Anatomy of Plants,* published in 1682, remained the most authoritative work in its field for more than 150 years.

The inventor: Nehemiah Grew

1641 Born in Atherstone, Warwickshire, the son of a cleric and schoolmaster

1661 Graduates with a B.A. from Cambridge University, England. Unable to continue studying at Cambridge because of his religious non-conformity, he then studies for an M.D. at Leyden University, Netherlands

1672 Moves from Coventry to London to study plant anatomy after being persuaded to do so by Fellows of The Royal Society, who raise £50 by subscription to encourage him to move. Publishes *An Anatomy of Vegetables Begun*

1677 Becomes secretary of The Royal Society

1681 Publishes *Comparative Anatomy of the Stomach and Guts*

1682 Publishes *Anatomy of Plants*, which is to remain the standard work on the subject for 150 years

1697 Publishes *A Treatise of the Nature and Use of the Bitter Purging Salt Contained in Epsom and Other Such Waters*

1698 Patents a method of preparing Epsom salts

1701 Publishes a religious and philosophical treatise, *Cosmologia Sacra*

1712 Dies on March 25 in London

Did you know?

As well as for its salts, the town of Epsom is also famous for horse racing. The Epsom Races were established in the 17th century and took place over four days in May; the most famous of the Epsom Races are the Derby (named after its cofounder, the Earl of Derby) and the Oaks (named after the Earl of Derby's estate near Epsom).

An 18th-century mock-epitaph criticizing Cheltenham spa waters read:

Here lies I and my three daughters,
Died from drinking Cheltenham waters;
Had we but stuck to Epsom salts,
We wouldn't be lying in these cold vaults.

Opposite top A waitress serves spa water to motorcycle passengers outside Harrogate's Royal Pump Room, England (1921)
Below Wounded soldiers drink spa water in England (c. 1915)

Diabetes was recognized as a disease by the ancient Egyptians and named by the ancient Greeks, but it was not until 1921 that its cause was identified, and a treatment, insulin, was isolated by Canadian chemist Frederick Banting.

The inventor: Frederick Banting

1891 Born Frederick Grant Banting in Alliston, Ontario, the son of a farmer. He graduates in medicine from the University of Toronto and begins work as a general practitioner

1920 Begins research into identifying the pancreatic

secretion responsible for controlling blood sugar levels

1921 With Charles Herbert Best and James Bertram Collip (both Canada), Banting succeeds in isolating a pure extract of insulin

INSULIN

The fact that diabetes is linked to the pancreas was confirmed in 1889 when two German researchers, Joseph von Mering and Oskar Minkowski, removed the pancreas from dogs, and the dogs then developed diabetes. Scientists concluded that the pancreas must produce a chemical messenger that controls blood sugar levels, but they were a long way from identifying the chemical in question.

Then, in 1920, Frederick Banting (Canada) came up with an idea that seemed too simple to be true. Because the pancreas secreted a protein-destroying enzyme, Banting thought that it might be destroying the very protein that scientists had been searching for, and that if the enzyme-secreting ducts were removed, the remainder of the pancreas might yield the elusive chemical messenger. After some persuasion, John MacLeod (Scotland), professor of physiology at the University of Toronto, accepted Banting into his department, where Banting, assisted by research student Charles Best (Canada), proved his theory correct. Then, in July 1921, with the help of biochemist Bertram Collip (Canada), they managed to isolate a pure extract of the vital protein, which they called insulin, and the three of them patented: "A method of preparing extracts of pancreas, suitable for administration to the human subject."

The first patient to be given insulin was 14-year-old Leonard Thompson (Canada), who was treated at Toronto General Hospital on January 11, 1922, by Walter A. Campbell and Alma A. Fletcher (both Canada)—Thompson had been expected to die, but he survived and went on to lead a relatively normal life with daily doses of insulin. The discovery and production of insulin was such an important medical breakthrough that Banting and MacLeod were jointly awarded the 1923 Nobel Prize for Physiology or Medicine. Banting was outraged that Best had not been recognized by the Nobel Prize committee and gave half of his prize money to Best, while MacLeod gave half of his to Collip.

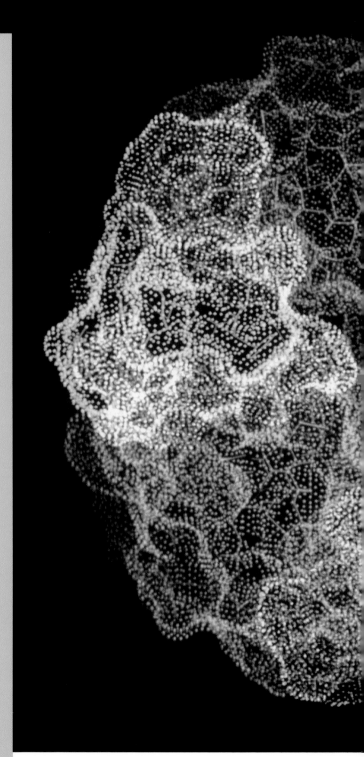

1922 A human patient is administered insulin for the first time; the treatment is successful

1923 Awarded the Nobel Prize for Physiology or Medicine jointly with John James Rickard MacLeod (Scotland). Becomes a professor at the University of Toronto

1924 Establishes the Banting Research Foundation

1930 Establishes the Banting Institute in Toronto

1934 Receives a knighthood

1941 Killed in an air crash in Newfoundland

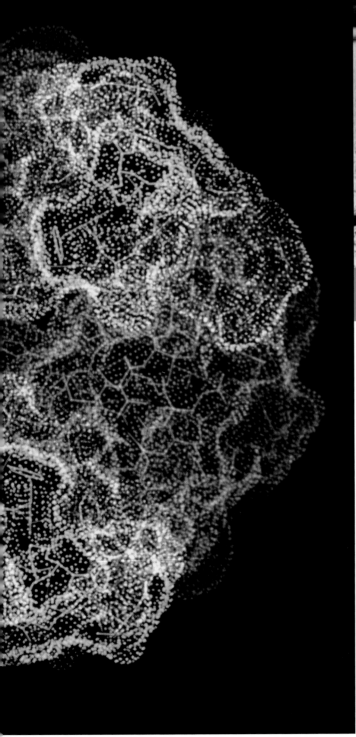

Left Computer-generated image of an insulin molecule
Above Charles Best and Frederick Banting on the roof of the medical building at Toronto University, with one of the first diabetic dogs to receive insulin (1921)

Did you know?

Insulin takes its name from the Latin *insula*, meaning "island," because it is secreted by groups of pancreatic cells known as the islets of Langerhans.

In 2003 scientists at Texas A&M University and Penn State Chemical Engineering announced the invention of a "smart tattoo," which is being developed to warn diabetics when their glucose levels are low. The tattoo consists of fluorescent molecules that are injected under the skin, and that fluoresce when glucose levels drop. When the smart tattoo is perfected, diabetics will be able to read the fluorescence levels (thereby monitoring glucose levels) using a watch.

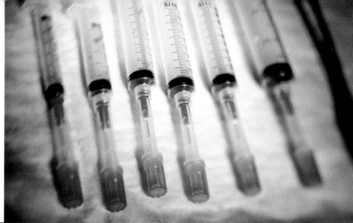

American global medical technology company Becton Dickinson has a long and illustrious history of developing syringes and injection devices. One of the most important was the disposable syringe, invented in 1954.

DISPOSABLE SYRINGE

The need for cleanliness in medical practice has been recognized since the pioneering work in antisepsis carried out by Ignaz Semmelweis (Hungary) and Joseph Lister (England) in the mid-19th century. Syringes were no exception, but they proved difficult to clean and sterilize properly. The perfect solution was not to reuse them at all, making the disposable syringe a huge boon to medicine.

Becton Dickinson was founded in 1897 by Maxwell W. Becton and Fairleigh S. Dickinson (both U.S.), who introduced their first injection device in October of that year. At that time, most syringes comprised a steel casing with a glass reservoir, but Becton Dickinson changed that by popularizing all-glass syringes after acquiring a half interest in the patent rights to a glass syringe that had been developed in 1869 by H. Wulfing Luer (France). In 1924 Becton Dickinson designed a syringe specifically for the injection of artificial insulin, which had first been synthesized for human use in 1922, and the company remains the world's largest manufacturer of insulin injection devices to this day.

In 1954 Becton Dickinson again came to the fore when the company produced the world's first disposable syringe in order to facilitate the mass administration of Jonas Salk's (U.S.) famous polio vaccine—seven million doses of the Salk vaccine had been injected by the end of 1955. These early glass disposable syringes did not prove cost-effective, but Becton Dickinson continued to develop the idea and in 1961 introduced the Plastipak plastic disposable syringe. Becton Dickinson became a public company in 1962 in order to fund the mass production of disposable syringes and other disposable medical devices—the change from glass to plastic saw production rise from 150,000 reusable glass syringes a week in the 1950s to 14 million disposables a week by the end of the century.

See also: Insulin, page 204

Becton Dickinson Inc. time line

1897 Company founded. Introduces its first injection device in October

1898 Acquires a half share of the patent for the first all-glass syringe

1906 Establishes the first factory in the U.S. specifically built for the production of thermometers, hypodermic, needles and syringes

1924 Introduces a syringe designed specifically for the delivery of insulin

1925 Fairleigh S. Dickinson (U.S.) invents and patents the Luer-Lok tip, which securely attaches a hypodermic needle to a syringe

1947 Introduces the first evacuated blood collection tube, known as the Vacutainer system (patent granted 1949)

1954 Produces the first disposable syringe

1961 Introduces the Plastipak disposable plastic syringe

Above opposite Disposable syringes make injections less expensive, but no less unpleasant **Below** A doctor administers an inoculation at the Wednesday afternoon children's clinic at Bethel Lutheran church, Philadelphia (1955)

1962 Goes public to fund the mass production of disposable medical devices

1970 Introduces the first disposable, self-contained insulin syringe with permanently attached needle

1973 Invents the fluorescence-activated cell sorter, commonly used today in cancer and HIV/AIDS research

1975 Invents and patents the Hypak prefilled syringe for injecting the blood-thinning agent heparin

1982 Becomes the first injection device manufacturer to introduce specially designed plastic needle-disposal containers

1988 Introduces the Safety-Lok syringe—the first syringe with an integrated needle safety feature. Begins field trials for the SoloShot syringe, which automatically locks after a single use

1989 Introduces plastic (as opposed to glass) blood collection tubes

1991 Introduces the InterLink IV Access Cannula, a needleless cannula for IV use. Introduces the SoloShot syringe, the first reuse prevention device to be released for commercial sale

1992 Introduces the Safety-Lok blood collection set. Introduces the Saf-T-Intima IV catheter

1995 Introduces the Autoguard IV catheter with a push-button retracting needle

1997 Introduces the SafetyGlide hypodermic needle,

featuring single-handed activation of the safety shield

1998 Introduces the Uniject prefilled injection device, designed for immunizations in the most remote locations of the developing world

2002 Introduces the Integra syringe, the first retracting safety syringe with a detachable needle

2003 Introduces the first blood collection set with a retracting needle

The adhesive bandage was invented independently on both sides of the Atlantic during the same decade. Johnson & Johnson was first off the mark in 1921 with Band-Aid, and Britain followed in 1928 with Smith & Nephew's Elastoplast.

ADHESIVE BANDAGE

The world's first pre-prepared adhesive bandage was Johnson & Johnson's J&J Band-Aid Adhesive Bandage, invented by employee Earle E. Dickinson (U.S.) in 1920 and launched commercially in 1921. Johnson & Johnson (U.S.) was founded in 1885 as a surgical dressings manufacturer and had been producing adhesive surgical tape, gauze, and other dressings for years before 1920—but the adhesive bandage did not arise from extensive research and development; it was the result of a piece of problem-solving by Dickinson, whose wife, Josephine, was particularly accident prone.

Dressing Josephine's wounds was becoming a time-consuming process for Dickinson, so he decided to make a ready-to-use bandage from the company's existing adhesive tape and gauze. He rolled out a length of surgical tape, placed short strips of gauze dressing on it, covered it with crinoline to prevent it sticking to itself, and then rolled it up again. Now all he (or she) had to do was cut the ready-made dressings off the roll when they were needed. At first, Dickinson's invention did not catch on, but in 1924 Johnson & Johnson began producing it in precut strips rather than a roll, and Band-Aid soon became an indispensable household item.

In 1928 another form of adhesive bandage was invented in Britain, originally used as a treatment for varicose ulcers. In 1856 Thomas Smith (England) established a business as an analytical and pharmaceutical chemist in Hull, England. Forty years later his nephew Horatio Nelson Smith was made a partner, and in 1928 Horatio, now in sole charge of the company, invented a new adhesive elastic dressing. It was slow to catch on until a surgeon published an article saying that the new dressing was of benefit to patients suffering from varicose ulcers, and Smith cannily circulated the article among the medical profession. Before long the new dressing was one of Smith & Nephew's best-selling products for general use, and the description elastic plaster bandage provided the trade name Elastoplast.

Did you know?

Johnson & Johnson began producing its famous baby powder in 1890 as a direct result of its work in surgical dressings. A doctor wrote to the company to say that one of his patients was suffering from a skin complaint as a result of using adhesive surgical tape, so the company began including a tin of talcum powder with the tape. Soon the talc was as popular as the tape, and Johnson's Baby Powder was born.

Before she achieved fame as a film actress, Brooke Shields appeared in a Band-Aid commercial in 1971.

The inventors: Johnson & Johnson and Smith & Nephew

1856 Thomas James Smith (England) opens a pharmacy in Hull, England, calling himself an analytical and pharmaceutical chemist

1885 Thomas Wood Johnson and James Wood Johnson (both U.S.) establish Johnson & Johnson, manufacturer of surgical dressings, in New Brunswick, New Jersey

1887 Johnson & Johnson is incorporated and, inspired by the work of Sir Joseph Lister (England) in antiseptics, begins developing sterile medical dressings

1896 Smith's nephew Horatio Nelson Smith becomes a partner in the company, which is renamed T.J. Smith & Nephew. Thomas Smith dies shortly afterward. Horatio has previously worked in textiles, giving him the expertise to begin manufacturing bandages

1907 Smith & Nephew becomes a limited company

1920 Johnson & Johnson employee Earle E. Dickinson (U.S.) invents a pre-prepared adhesive dressing

1921 Johnson & Johnson launches Dickinson's invention as Band-Aid

1928 Horatio Smith oversees the invention of Elastoplast

Opposite Everybody needs a Band-Aid every so often—even world-famous soccer stars like David Beckham (2003)
Left Band-Aids can also be used to prevent injuries, such as blisters from high-heel straps

Since Vaseline is such a common item to be found in any medicine cabinet, it is hard to imagine that someone had to invent or patent it—but that is exactly what Robert Chesebrough did in 1872, after discovering the wonders of petroleum jelly.

VASELINE

Robert Chesebrough (England–U.S.) was a 19th-century chemist who made his living from selling products derived from the oil of sperm whales. In 1859 he heard about an oil strike in Pennsylvania and went to investigate this new source of oil and, potentially, oil products. What he found was to make him a very rich man. While visiting the oil field he was told about a jellylike substance called rod wax, which was a nuisance to the riggers because it clogged up the drills; they would simply scrape it off the drilling equipment and throw it away. He also heard rumors that this rod wax had miraculous healing powers, seemingly confirmed by the fact that the riggers applied it to cuts and burns, which then healed much more quickly than usual.

Chesebrough took some of the black, gooey wax back to Brooklyn, where he applied his skills as a chemist and discovered that it was actually petroleum jelly, which in its pure form was clear and odorless. It is said that he tested it by deliberately cutting and burning himself and then smearing the jelly on the injuries—miraculously, the jelly helped them to heal. Chesebrough patented his method of "treating hydrocarbon oils etc.," and named the resulting jelly Vaseline, a trade name that takes its name from the German *wasser*, meaning "water," and the Greek *elaion,* meaning "oil." He then traveled around New York cutting and burning himself, treating himself with Vaseline, and then selling his "water-oil" from the back of a wagon—before long Vaseline had become an international success and Chesebrough had made his fortune.

Chesebrough's patent stated that Vaseline could be used for a number of purposes from treating leather to a pomade for the hair, and it has since been used for everything from greasing machinery to removing stains from furniture. It has also remained a popular cure-all even after its miraculous healing powers were explained by the fact that it worked not by some mysterious magic, but by simply sealing wounds from infection.

Below Tubs of Vaseline in soft focus, achieved by smearing Vaseline over

the camera lens (2003) **Above** Workers drilling for oil on an offshore oil rig, California

The inventor: Robert A. Chesebrough

1837 Born Robert Augustus Chesebrough in England and later immigrates to the U.S.

1859 Travels to Pennsylvania to investigate an oil strike

1872 Granted a U.S. patent on June 4 for: "Treating hydrocarbon oils etc."

1874 Granted a U.K. patent for: "Treating hydrocarbon oils etc."

1878 Registers the name Vaseline as a trademark

1933 Dies aged 96

Did you know?

Robert Chesebrough, who lived to the age of 96, is said to have remained healthy by eating a daily spoonful of Vaseline.

Kerosene takes its name from the Greek *keros,* meaning "wax."

The original name of English rock band Elastica was Vaseline.

The first artificial heart to serve as a permanent replacement for a human heart was the Jarvik-7, invented by Robert K. Jarvik in 1977 and implanted by surgeon William de Vries in the body of 61-year-old dentist Barney Clarke in 1982.

Did you know?
Pioneering work in artificial hearts was carried out by a team of scientists led by Willem Kolff (Netherlands–U.S.), whose heart pumps were tested in animals as early as 1957. Kolff had previously invented the kidney dialysis machine in 1943, and immigrated to the U.S. in 1950.

ARTIFICIAL HEART

Heart surgery progressed very quickly after the first human heart transplant in 1967, and Robert Jarvik (U.S.) knew that he was not alone in thinking that it might be possible to replace a defective human heart with an artificial one—his U.S. patent (filed 1977, granted 1979) discusses much of the work that had already been done, which included more than 100 U.S. patents relating to artificial heart pumps. Despite all this prior work, Jarvik can be described as the inventor of the artificial heart because he designed the first such pump to permanently replace a human heart.

When Jarvik was 18, his father underwent open-heart surgery, a traumatic event that prompted Jarvik to study medicine; he graduated from the University of Utah in 1976 with an M.D. The following year he filed his patent for an artificial heart, and in 1982 surgeon William de Vries (U.S.), of the University of Utah Medical Center, implanted a Jarvik-7 in Barney Clarke (U.S.), who survived for 112 days. Jarvik's 1977 patent suggests that at that stage the artificial heart was intended as a temporary rather than a permanent replacement, but five years later, when it was implanted in Clarke, the Jarvik-7 was capable of being just that, with two separate pumps acting like the ventricles of the heart.

Although the Jarvik-7 had prolonged the lives of more than 70 patients by the end of the 1980s, there were problems, including an increased risk of blood clots, strokes, haemorrhaging and secondary infections. Jarvik continued researching to improve the heart, and at the turn of the millennium he was working on the Jarvik-2000, designed to be connected to a coin-sized controller screwed into the skull behind the ear and powered by a battery pack worn on the belt. To date the longest surviving recipient of an artificial heart is William Schroeder (U.S.), who received a Jarvik-7 on November 25, 1984, implanted by William de Vries, and who survived a total of 620 days until August 7, 1986.

Left Robert Jarvik with Jarvik-7 artificial heart (1984) **Above** Jarvik 2000 artificial heart (2000)

The inventor: Robert K. Jarvik

1946 Born Robert Koffler Jarvik on May 11 in Michigan

1964 While Jarvik is studying at the University of Utah, his father undergoes open-heart surgery, prompting him to change courses and study medicine

1976 Graduates in medicine from the University of Utah

1977 Files a patent for: "Total Artificial Hearts And Cardiac Assist Devices Powered And Controlled By Reversible Electrohydraulic Energy Converters" (granted 1979)

1989 Granted patents for an artificial ventricle and for a: "Multiple-Electrode Intracochlear Device"

1990 Granted a patent for: "Prosthetic Compliance Devices"

1991 Granted a patent for: "Intraventricular Artificial Hearts And Methods Of Their Surgical Implantation And Use"

1992 Granted a second patent for: "Intraventricular Artificial Hearts And Methods Of Their Surgical Implantation And Use"

1994 Granted a patent for: "Cannula Pumps For Temporary Cardiac Support And Methods Of Their Application And Use"

2000 Begins work on the Jarvik-2000 heart pump

Did you know?
Although they have helped many people who had no other option, artificial hearts have still not improved enough to match the real thing. The longest surviving recipient of a human heart was Dirk van Zyl (South Africa), who lived for 23 years and 57 days after his operation in 1971, compared with a life expectancy of less than 2 years for the recipient of an artificial heart.

The electric hearing aid was invented by Miller Reese Hutchinson in 1901, but in 1979 an even more exciting advance, which sounds like something from a sci-fi novel, was invented by Graeme Clark— the bionic ear.

BIONIC EAR

For centuries, people have tried to aid their hearing by cupping a hand around their ear or using devices such as ear trumpets to amplify sound. Then, in 1901, Miller Reese Hutchinson (U.S.) invented the first electric hearing aid, which he called the Acousticon. It was a cumbersome device weighing 15 pounds (7 kilograms), with a telephone-type receiver that was held to the ear, but it was the first step toward much smaller modern hearing aids. The first truly portable hearing aid was the Amplivox, which weighed just 2 pounds (1 kilogram) and was invented in 1935 by A. Edwin Stevens (England). The transistorized hearing aid followed in 1952, invented by Sonotone Corp. (U.S.), heralding an ever accelerating period of development that saw hearing aids small enough to be worn completely within the ear, and digital hearing aids in which individual frequencies could be separately amplified.

But for all the help they have brought to millions of people, hearing aids are just that—aids to a defective but functioning sense of hearing. Inspired by the plight of his totally deaf father, Professor Graeme Clark (Australia) developed an invention that was to bring hearing to the profoundly or totally deaf: a bionic ear. "Bionic" sounds like a word invented by the makers of *The Six Million Dollar Man*, but in fact it is a scientific term with two meanings, one of which refers to the replacement of damaged or defective parts of the body by electronic devices that work with the body. Clark's invention was a cochlear implant that enabled profoundly and totally deaf children and adults to hear, by electrically stimulating the hearing nerves in the inner ear (cochlea)—essentially, an electronic ear that interacted directly with the body.

The first bionic ear was implanted in a patient's mastoid bone (part of the cochlea) in August 1978, and since then Clark has continued to develop and improve his cochlear implant. He was awarded the Order of Australia for his contribution to medicine in 1983 and elected an Honorary Fellow of The Royal Society of Medicine in 2003.

The inventor: Graeme Clark

1935 Born Graeme Milbourne Clark in Melbourne, Australia

1957 Graduates from the University of Sydney, Australia, with bachelor of medicine and bachelor of surgery degrees

1967 Begins research at the University of Sydney into the possibilities of an electronic cochlear implant

1968 Awarded a master of surgery degree by the University of Sydney, his thesis being: "The principles of the structural support of the nose and their application to nasal and septal surgery"

1969 Awarded a Ph.D. by the University of Sydney, his thesis being: "Middle ear and neural mechanisms in hearing and in the management of deafness"

1970 Becomes research leader of the Department of Otolaryngology at the University of Melbourne

1978 Performs his first cochlear implant

1981 The University of Melbourne, the Australian government, and medical equipment manufacturer Nucleus Ltd. begin to develop a commercially practical cochlear implant

1982 Cochlear Ltd. is established to oversee commercial operations

1983 Clark is awarded the Order of Australia for his contribution to medicine

1984 Clark is appointed director of the newly established Bionic Ear Institute in Melbourne

1998 Clark is appointed a Fellow of the Australian Academy of Science and of the Australian Academy of Technological Sciences and Engineering

2003 Elected an Honorary Fellow of The Royal Society of Medicine. For the second consecutive year, Cochlear is named the most innovative company in Australia by the Intellectual Property Research Institute of Australia and the Melbourne Institute of Applied Economic and Social Research

Below Angela Sinclair demonstrating a Chase silver ear trumpet, made in 1880 and used by Queen Victoria, at the "Escape From Deafness" exhibition in Park Lane House, Park Lane, London (1958)
Right A model showing a cochlear implant (in yellow), or bionic ear, positioned in the inner ear (1999)

Lasers are now put to use in applications such as telecommunications, weapons systems, metal cutting and CD players, but the first lasers were used as medical tools. The first practical laser was built by physicist Theodore Maiman in 1960.

LASER

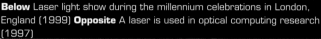

Did you know?
Albert Einstein had three nationalities. He was born in Germany, took Swiss nationality in 1901 (where he worked in the Swiss patent office from 1902–05), and became a U.S. citizen in 1940.

Below Laser light show during the millennium celebrations in London, England (1999) **Opposite** A laser is used in optical computing research (1997)

A laser is a device that produces a concentrated beam of light that can be very accurately controlled and used for a variety of purposes. It works by stimulating the emission of electromagnetic radiation, hence the name light amplification by stimulated emission of radiation, derived from the earlier maser, which amplified microwaves rather than visible light.

As early as 1916, Albert Einstein (Germany–Switzerland–U.S.) stated that the stimulated emission of radiation was theoretically possible. The first patent for "a method for the amplification of electromagnetic radiation" was filed by three Soviet scientists in 1951 (granted 1959), but there was no suitable amplifying medium to make their invention a practical reality. Meanwhile, Charles Hard Townes, James P. Gordon, and Herbert Zeigler (all U.S.) invented a working maser in 1954, the first practical demonstration of Einstein's theory.

Then, in 1958, Townes and his brother-in-law Arthur Leonard Schawlow filed a patent (granted 1960) relating to "masers and maser communication system"; this was actually the first patent for a laser, because it related to what they called an "optical maser," an invention that later became known as a laser. But the first *working* laser was built in 1960 by Theodore Harold Maiman (U.S.), who used a ruby as the amplifying medium while most other scientists were researching the use of various gases. His laser was first used in 1964 as a "laser lancet" for retinal surgery.

Maiman's laser was based on the Schawlow–Townes patent, but a legal battle later ensued when Gordon Gould (U.S.) claimed to have invented the laser three years earlier in 1957. He had kept copious notes, verified by a lawyer, but had not applied for a patent because he mistakenly thought that he must first build a working prototype. Gould lost the first round of court cases, but was later granted several key patents giving him a strong claim to be named as inventor of the laser.

The inventors: Schawlow, Townes, Gould, & Maiman

1915 Charles Hard Townes is born on July 28 in Greenville, South Carolina, a lawyer's son

1920 Gordon Gould is born in Pittsburgh, Pennsylvania, the son of a magazine editor

1921 Arthur Leonard Schawlow is born on May 5 in Mount Vernon, New York, the son of an insurance agent

1927 Theodore Maiman is born on July 11 in Los Angeles, the son of an electrical engineer

1939 Townes graduates from Cal Tech with a physics Ph.D.

1942–45 Townes designs radar bombing systems for Bell Laboratories (U.S.)

1948 Townes becomes associate director of physics at Columbia University

1949 Schawlow graduates from University of Toronto with a Ph.D. in physics and spends two years as a postgraduate at Columbia University, where he meets Townes and marries his youngest sister

1954 Townes, Gordon, and Herbert Zeigler (U.S.) build the first maser

1955 Maiman graduates from Stanford University with a Ph.D.

1957 Gould, while studying for a Ph.D. at Columbia University, conceives the idea and name of the laser, but delays his patent application until 1959

1958 Townes and Schawlow file the first patent for a laser (granted 1960)

1960 Maiman announces the first practical laser

1964 Townes shares the Nobel Prize for Physics with Nikolai Bosov and Aleksandr Prokhorov (both U.S.S.R.) for their work in masers and lasers

1970–79 Various court cases brought in respect of Gould's patents result in patent protection for Gould's laser and the recognition of Gould as an inventor of the laser independently of Townes and Schawlow

1981 Schawlow shares the Nobel Prize for Physics with Nicolaas Bloembergen (Netherlands–U.S.) & Kai Siegbahn (Sweden) for their "contribution to the development of laser spectroscopy"

The invention of the Dentron Biogun by Jonathan Copus was a revolution in infection control—a quick, painless, noninvasive, drug-free treatment with no known side effects that has been widely used in chiropody, dentistry, and dermatology.

BIOGUN

The Biogun is effectively an electronic antibiotic. It kills the bacteria, fungi, and viruses that cause infection by bombarding them with a concentrated stream of negatively charged air particles—mainly superoxide radical anions—and all the patient feels is a gentle breeze of air streaming from the tip of a "wand" known as an ion emitter.

This revolutionary device was invented in 1994 by Jonathan Copus (England, b. Scotland) during a bout of athlete's foot: "I got fed up with squelching around with socks full of creams and powders. I read somewhere that athlete's foot was caused by a fungus and I read somewhere else that fungi and bacteria didn't like a negatively charged atmosphere. So I thought, 'What would happen if you produced a very concentrated, negatively charged atmosphere and shone it on the spot?' I built a very crude prototype, which produced 7,500 volts and ended in a sewing needle producing a concentrated stream of negatively charged particles. I put that between my toes for a fortnight and at the end of that time, lo, the athlete's foot was no more!"

Electronic antibiotics may sound like a highly unnatural way of treating infection, but in fact superoxide is one of the immune system's own weapons against infection—and while the voltage is high, the current is an extremely low 100 microamperes, making the Biogun perfectly safe. It found favor first with dentists, then with chiropodists, and its use has since spread to general medicine. Scientifically speaking, the stream of superoxide anions "acts as a nucleophile on the phospholipid bilayer, causing a de-esterification of the fatty acids and weakening the cell membrane, leading to lysis." In other words, the bacteria burst open and die. The result is that any number of infections on or near the surface of the skin can be treated without drugs, pain, or side effects, with proven success against infections ranging from mouth ulcers and gingivitis to verrucae (warts) and, of course, athlete's foot.

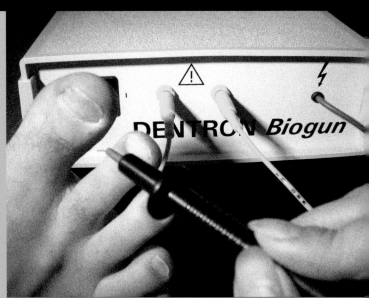

Above The Dentron Biogun uses a concentrated stream of negatively charged air particles to treat athlete's foot, among other ailments **[Opposite]**

The inventor: Jonathan Copus

1944 Born on September 6 in Dundee, Scotland

1949–66 Educated at High School of Dundee, Trinity School of John Whitgift, Croydon, and Brasenose College, Oxford

1966–89 Works as a teacher in London, as a parson in Woking, and as a broadcaster and journalist for the BBC in Southampton and London

1989 Invents an electronic painkiller for dentists, which works by producing specially shaped electronic signals that "switch off" pain messages before they reach the brain at the same time as stimulating the release of endorphins, the body's natural painkillers. Founds Dentron Ltd.

1990–93 Invents the Painaway electronic painkiller for chiropodists. Invents the world's first frequency-modulated acupuncture stimulator

1994 Invents the Biogun, which he patents on April 6 as "destruction of micro-organisms." Becomes a member of the Institute of Patentees and Inventors (Britain)

1994–2002 Wins four SMART awards to develop electronic microbicidal technology in various fields, the latest for treating diabetic foot ulcers

Incubators for hatching chickens had been known for many years before the first medical incubator for premature human babies was invented by Stéphane Tarnier and designed by zoologist Odile Martin, who patented it in 1880.

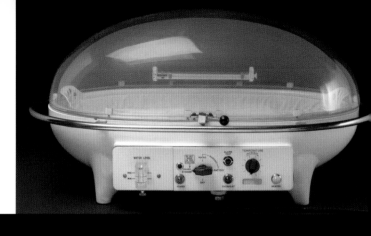

INCUBATOR

In 1824, when she was five and still a princess, Britain's Queen Victoria was presented with a gift of chicks that had been hatched in an incubator. Despite this high-profile exposure of the concept of artificial incubation, it was to be more than half a century before obstetric surgeon Stéphane Tarnier (France) became the first person to think of using an incubator for the care of premature human babies.

Tarnier graduated in medicine in 1850 and subsequently began working at the Maternité Hospital in Paris, intending to gain some obstetric experience before going into general practice. However, his research into puerperal fever, which was killing many women giving birth at the Maternité, led him to specialize in obstetrics instead, and in 1867 he became surgeon-in-chief at the hospital.

In 1878 Tarnier was visiting Odile Martin (France), the director of the Paris Zoo, when he saw an incubator that had been designed for hatching exotic birds. Tarnier asked Martin to adapt the apparatus for human babies, and on April 9, 1880, Martin filed a patent for: "Improvements in incubators, partly applicable for other purposes." Improvements over the avian incubator included double walls filled with sawdust for insulation, a 126 pt. (60 l) reservoir of hot water at the bottom to warm a separate upper chamber, and doors opening at the sides of the upper chamber for access to the babies. Warmed cradles had already been used in some hospitals, but this was the first closed-top hospital incubator, the "lid" being a double-glazed panel that allowed the babies, and the all-important thermometer, to be monitored.

Tarnier put this incubator into use at the Maternité Hospital, where the death rate for babies with low birth weights quickly fell by almost half, and in 1893 his protégé, Pierre Budin (France), who is sometimes erroneously credited with inventing the incubator, established the world's first specialist unit for premature babies.

See also: Thermostat, page 16

The inventor: Stéphane Tarnier

1828 Born Étienne Stéphane Tarnier on April 29 in Aiserey, France, the son of a country doctor

1845 Graduates from the Lycée of Dijon and goes to Paris to study medicine

1850 Qualifies in medicine

1856 Enters the Maternité Hospital to gain obstetric experience, intending to then go into general practice

1857 Presents a dissertation to the Académie on the subject of puerperal fever, which is killing many women giving birth in the Maternité

Did you know?

Six incubators were exhibited at the Berlin Exposition of 1896, for which Pierre Budin's assistant, Martin Couney (France), "borrowed" six premature babies from a local hospital to demonstrate the incubators in action. The incubators proved to be one of the most popular exhibits at the exposition, but when Couney repeated his stunt, the medical journal *The Lancet* asked: "Is it in keeping with the dignity of science that incubators and living babies should be exhibited amidst the aunt-sallies, the merry-go-rounds, the five-legged mule, the wild animals, the clowns, the penny peep-shows, and amidst the glare and noise of a vulgar fair?"

Opposite top Portable neonatal incubator by Oxygenaire (1950–65) **Below** Incubation units in use at the Maternité Hospital on Port-Royal Boulevard in Paris, where they had been in operation for the care of weak and premature babies since 1881 (c. 1884)

221

1867 Becomes surgeon in chief at the Maternité

1867–89 Becomes the first French surgeon to implement Semmelweiss and Lister's principle of antisepsis and reduces puerperal fever deaths from 9.3 percent of births to just 0.7 percent *See also*

Disposable syringe, page 206

1877 Invents the axis traction forceps, a great improvement on earlier forceps used for delivery of babies

1883 Invents an obstetric instrument known as a basiotribe

1886 Made honorary LL.D. of Edinburgh University, Scotland

1889 Becomes clinical professor of obstetrics at the University of Paris

1891 Elected president of the French Academy of Medicine

1894 Publishes a 40-year study into puerperal fever

1897 Suffers a stroke on the day of his retirement and dies on November 22, aged 69

The discovery of x-ray photography in 1895 was a huge boon to medical diagnosis. Seventy-three years later, Godfrey Hounsfield put x-rays to even more impressive use when he invented computer-aided tomography (CAT) in 1968.

CAT SCANNER

The discovery of x-rays by Wilhelm Konrad von Roentgen (Germany) in 1895 had such profound implications for medical diagnosis that Roentgen was awarded the Nobel Prize for Physics when the prizes were inaugurated in 1901. Although x-rays provided a revolution in diagnostic techniques, their use was limited; being two-dimensional, they could not show the depth of an injury within the body. During the 1960s Godfrey Hounsfield (England) was researching a way of enabling computers to recognize patterns so that they could "read" images. Having lectured in the principles of radar during the Second World War, he combined the ideas of pattern recognition and radar rangefinding as the basis of what became known as computer-aided tomography (CAT) or computerized tomography (CT).

At first CAT was developed to scan the brain—several x-rays would be taken of different cross sections of the brain and then combined by the computer to create a three-dimensional image. Not only was the CAT image clearer than a single x-ray, but the process was far safer than the conventional method, which involved introducing chemicals to the brain in order to be able to create a meaningful x-ray image. Hounsfield filed a patent for his idea on August 23, 1968, but although the principle worked, computer technology was not far enough advanced to make a practicable CAT scanner: Early prototypes took over four hours to scan an object and then produce a computerized image.

The first practicable CAT scanner was built by EMI in 1971 for scanning the brain. By 1975 full-body scanners were available and CAT was celebrated as the biggest advance in diagnostic techniques since the discovery of x-rays 73 years earlier. The Nobel Foundation was in no doubt of its worth, and Hounsfield shared the 1979 Nobel Prize for Physiology or Medicine with Allan MacLeod Cormack (South Africa–U.S.), who had been carrying out theoretical research in the same field since the late 1950s.

See also: Dynamite, page 266; X-ray photography, page 196

Above Godfrey Hounsfield beside the prototype CAT scanner (1972)
Opposite Patient undergoing a CAT scan, seen from the scanner control room

The inventor: Sir Godfrey Hounsfield

1919 Born Godfrey Newbold Hounsfield on August 28 in Newark, Nottinghamshire, the son of a farmer. He later studies at the City & Guilds College, London, and Faraday House College of Electrical Engineering, London

1939–45 Works as a radar lecturer in the RAF during the Second World War

1951 Joins electronics company Thorn/EMI Ltd., specializing in computers

1968 Files a patent on August 23 for computer-aided tomography

1971 EMI builds the world's first CAT scanner

1972 Hounsfield becomes head of medical systems research at EMI

1975 Receives an honorary doctorate in medicine from City & Guilds College, London

1979 Receives the Nobel Prize for Physiology or Medicine jointly with Allan MacLeod Cormack (South Africa–U.S.)

1980 EMI sells its interest in Hounsfield's patent to General Electric Corp.

1981 Hounsfield is knighted

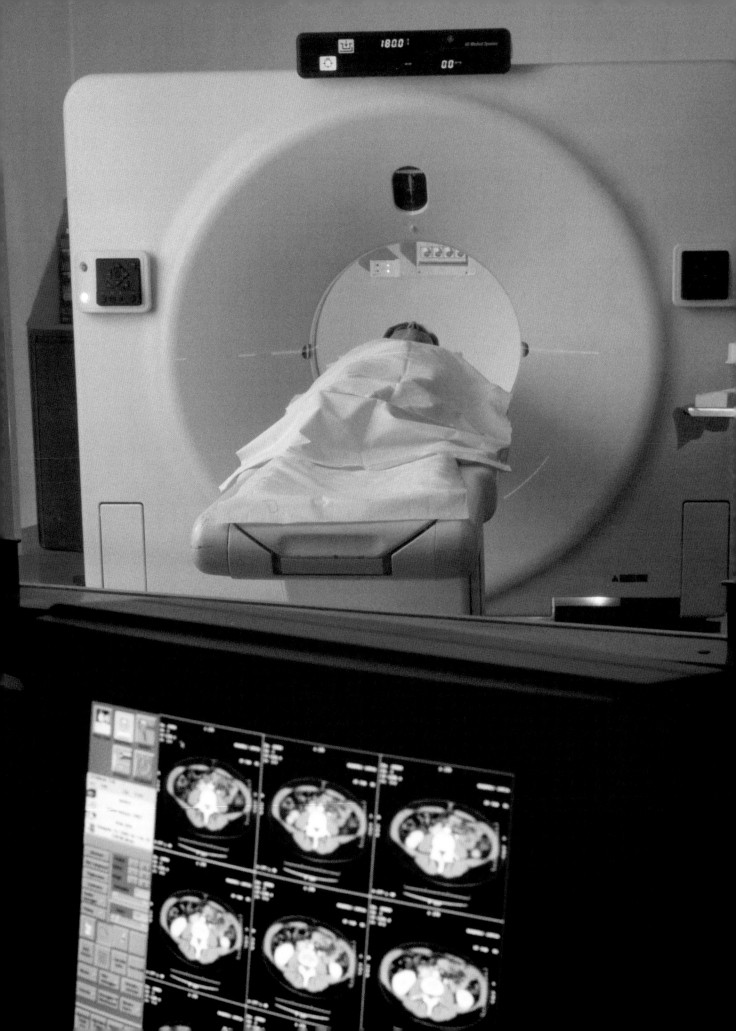

The idea of measuring intelligence was pioneered by Francis Galton in the late 19th century, but the first people to produce a rational series of tests and a scale of intelligence were Alfred Binet and Théodore Simon, in 1905.

INTELLIGENCE TEST

Psychologist Alfred Binet (France) did not approve of the concept of intelligence quotients (IQ), but, nonetheless, he is hailed as the inventor of the IQ test, having carried out research from 1904 to 1905 to assess the relative intelligence of French schoolchildren. Earlier, Francis Galton (England), a great supporter of his cousin Charles Darwin's theory of evolution, had championed the idea that intelligence was hereditary and could be measured. Galton propounded several controversial theories based on hereditary intelligence, including the concept of eugenics, the suggestion that there was an inherent difference in intelligence between races, and the idea that intelligence could be measured through physical features.

During the 1890s Binet began searching for a more scientific method of measuring intelligence, and for less controversial reasons than eugenics. He theorized that intelligence was a combination of comprehension, reasoning, attentiveness, memory, and imagination. Each of these attributes could be measured separately, and Binet began devising a series of tests that would arrive at a measurement of intellectual ability. In 1904 his research was officially sanctioned when the French government commissioned him to produce a system of testing the intelligence of schoolchildren so that the less intelligent could be referred to special schools.

Binet and his colleague Théodore Simon (France) devised a series of 30 tests, which they published in 1905 in *L'Année Psychologique,* together with a scale for registering the level of intelligence. The tests were intended to ignore the effects of nurture (i.e., things that the children had been taught) and to measure instead their natural mental ability, which Binet believed could alter with time. Binet disagreed with the suggestion that a fixed intelligence quotient could be arrived at by comparing "mental age" with actual age, but, nonetheless, his tests are the direct precursors of the modern IQ test.

Below Intelligence, perception, and dexterity are important factors in many jobs. Here, aptitude tests are conducted at the Chrysler Institute of Engineering (1956) **Right** Raven's Colored Progressive Matrices Test (1938)

The inventor: Alfred Binet

1857 Born on July 11 in Nice, France

1878 Abandons his law studies and moves to the Salpêtrière Hospital, Paris, to study under pathologist and neurologist Jean-Martin Charcot (France)

1892 Appointed director of physiological psychology at the Sorbonne, Paris

1896 Devises and applies his first intelligence tests

1898 Joined by Théodore Simon (France)

1905 Publishes *The Binet-Simon Test of Intelligence*

1911 Dies in Paris

Did you know?

The idea that an intelligence quotient could be extrapolated from intelligence tests such as Alfred Binet's by comparing mental age with actual age was suggested by psychologist and physiologist Wilhelm Max Wundt (Germany). The term intelligence quotient was popularized by psychologist Lewis Terman (U.S.) in his 1916 book *The Measurement of Intelligence*.

As well as propounding a number of controversial theories relating to inherited intelligence and eugenics, Francis Galton was also a renowned explorer and anthropologist, made significant advances in mapmaking and weather forecasting, and initiated the "nature versus nurture" debate that continues to this day.

AND ALSO...

Osteopathy/Chiropractic

The founder of osteopathy was Andrew Taylor Still (U.S.). He based his treatment on the principle "structure governs function" and began practicing in 1874; he founded the American School of Osteopathy in 1892. The first chiropractor was hypnotist and former osteopath Daniel David Palmer (U.S., b. Canada), who founded the Palmer School of Chiropractic in 1898.

Thermometer

The first thermometer was produced by Galileo Galilei (Italy) in 1592. Although it indicated variations in temperature, it could not measure them, and it was 1714 before Gabriel Daniel Fahrenheit (Germany, working in Holland) produced the first successful mercury thermometer and the first scale of temperature. Fahrenheit used human body temperature and the freezing point of saltwater as his fixed points, dividing the interval into 96 degrees. He later decided to use the boiling point of water as the upper fixed point and adjusted the scale so that boiling point was exactly 180 degrees above freezing point—the adjustment meant that instead of 96 degrees Fahrenheit, body temperature became 98.6 degrees Fahrenheit. When Anders Celsius (Sweden) devised his temperature scale in 1742, he described zero degrees as the boiling point of water, and 100 degrees as the freezing point. Only after his death was the scale inverted to its now familiar form. In 1948 the centigrade scale (literally meaning "100 degrees") was given the name Celsius.

Iron lung

Alexander Graham Bell (Scotland–U.S.) designed a prototype artificial respirator, or iron lung, in 1881, which he called a "vacuum jacket." The first practical iron lung was invented in 1927 by Philip Drinker (U.S.), who built a prototype using a sealed box with the air pressure controlled by a vacuum cleaner. A more sophisticated version was tested at Boston Hospital the following year, and the iron lung was produced for general use from 1931.

See also: Telephone, page 140

Homeopathy

The first person to practice homeopathy, and therefore its founder, if not its inventor, was physician Christian Frederick Samuel Hahnemann (Germany). Hahnemann, who based his ideas on the principle *similia similibus curantur* ("like cures like"), established that substances producing certain symptoms in healthy people will cure like symptoms in the sick and that remedies act not by quantity, but in infinitesimal doses. He began practicing homeopathy in 1796, but was persistently prosecuted for illegal medical practice, finally settling in Paris in 1835. By the time of his death in 1843, homeopathy had spread across Europe and North America, though still in conflict with conventional medicine.

CPR mannequin

The Actar 911 CPR mannequin was invented in 1989 by Dianne Croteau and her business partners Richard Brault and Jonathan Vinden (all Canada). It consisted of a head and torso on which students could practice CPR, or cardiopulmonary resuscitation. One big advantage of the Actar 911 over earlier, full-body mannequins was that it was cheap enough for classes to supply students with one each, where previously they had had to wait their turn to share a single mannequin.

Above top Too late for him. Students at the British School of Osteopathy in London learn the principles of osteopathic technique on a skeleton (1963) **Opposite** Firefighter practicing CPR with defibrillator in emergency-medicine class at Fire and Rescue Academy, Virginia (1997)

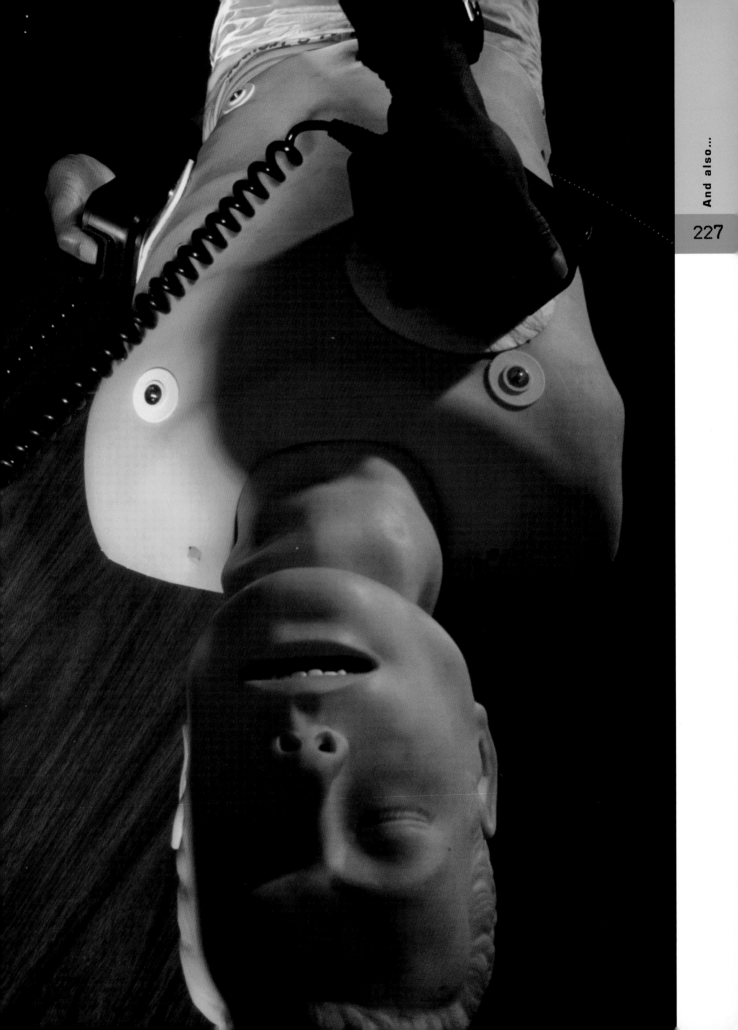

Chapter Seven

CUTTING EDGE

To most people in the 21st century, batteries are simply a convenient substitute if electricity is impractical or unavailable. But when the battery was invented in 1800 it was the first device capable of providing an electric current.

The inventor: Alessandro Volta

1745 Born Alessandro Giuseppe Anastasio Volta on February 18 in Como, northern Italy, and later educated at the local Jesuit College and the Seminario Benzi

1775 Appointed professor of physics at the Royal School,

Como. Invents the perpetual electrophorus

1777 Invents a hydrogen pistol fired by electricity

1778 Appointed professor of physics at the University of Pavia. Invents the electrical

BATTERY

Alessandro Volta (Italy) was an eminent physicist who made a number of significant breakthroughs in the field of electricity and is commemorated by the use of his name for the electrical unit of potential difference, the volt. His most famous invention, the battery, came about as the result of a disagreement with another Italian scientist, Luigi Galvani.

In 1775 Volta invented a "perpetual electrophorus": an early form of electrical induction machine that could be used to generate and store static electricity. It worked by rubbing cat fur across a rubber-coated metal plate to build up a static charge, and then transferring the charge to a Leyden jar; by repeating the process several times a large charge could be built up. (The Leyden jar was invented in 1745 at the University of Leyden, Netherlands, as a means of storing electrical charge.) Static electricity and the Leyden jar were nothing new, but what Volta had invented was a means of accumulating an electrical charge.

Volta's real breakthrough, however, came when he heard about the electrical experiments of anatomist Galvani. Galvani had noticed that dissected frogs' legs would twitch if they were part of a circuit involving two metals, and concluded that the muscles and nerves were the source of what he called "animal electricity." Volta disagreed, thinking that the current was more likely to be due to the connection between the two metals; he began experimenting with pairs of metals and discovered that he could generate an electric current if certain metals were submerged in acid. The result, in 1800, was Volta's greatest invention, the "voltaic pile." It was the world's first battery—a series of copper and zinc strips submerged in salt water that could provide an electric current rather than mere static electricity. The following year, Napoleon I (ruler of Italy since its conquest by France in 1796) bestowed the Legion of Honor on Volta, in recognition of the importance of his invention, and made him a count.

condenser, an electrical storage device later known as a capacitor (the later term coined by Volta in 1795)

1787 Invents the candle flame collector of atmospheric electricity

1795 Becomes rector of the University of Pavia

1800 Invents the "voltaic pile," or electrochemical battery. Publicly announces his invention on March 20 in the *Philosophical Transactions of The Royal Society.*

Demonstrates the battery to the Institut Française and Napoleon Bonaparte on November 18

1801 Awarded the Legion of Honor by Napoleon I, who also makes him a count

1827 Dies on March 5 in Como, aged 82

Below right Alessandro Giuseppe Antonio Anastasio Volta (1745-1827), inventor of the "voltaic pile"—the first electric battery
Below A beneficiary of Volta's brainchild

Did you know?

The volt is the SI (*Système International*) unit for potential difference. Other SI units named after people include: the amp, or ampère (electric current), after physicist André Marie Ampère (France); the watt (power), after engineer James Watt (Scotland); the joule (energy), after natural philosopher James Joule (England); the newton (force), after scientist Sir Isaac Newton (England); the henry (electrical inductance), after physicist Joseph Henry (U.S.); the farad (electrical capacitance), after chemist and physicist Michael Faraday (England); the coulomb (electrical charge), after physicist Charles Coulomb (France); and the tesla (magnetic flux density), after physicist and electrical engineer Nikola Tesla (Croatia–U.S.). *See also: AC induction motor, page 232*

The alternating current (AC) induction motor was invented by Nikola Tesla, who died with more than 100 U.S. patents to his name. He also proposed many unpatented ideas for radio, television, death rays, and even interplanetary communication.

Did you know?

Nikola Tesla invented a machine to generate high-voltage discharges and in 1900 set up a 198-foot-high (60-meter-high) tower in Denver to shoot these "lightning bolts" into the sky as a signal to beings on other planets. Naturally the neighbors complained about the disturbance, so he set up another tower next to a hospital for the deaf and mute.

AC INDUCTION MOTOR

The importance of the part played by physicist and electrical engineer Nikola Tesla (Croatia–U.S.) in the development of electricity as we know it today is often underestimated, because an alternating current (AC) induction motor does not sound like an epoch-making breakthrough. But what Tesla invented was the machinery and the principles that made it possible to supply electricity to people's homes.

Tesla's first invention, in 1881, was a telephone repeater, conceived while he was working for the Hungarian Telegraph Office. In 1882 he originated the idea that two sources of AC, out of phase, would create a rotating magnetic field, which in turn would make it possible to build an AC induction motor. Tesla built his first such motor in Strasbourg in 1883 before immigrating in 1884 to the U.S., where he worked for Thomas Edison (U.S.). The two men did not get on— Edison believed in direct current (DC), while Tesla could see that AC was the way forward. Tesla left Edison's company and in 1887 filed a single patent application covering several inventions, including single- and multi-phase AC motors, a distribution system, and a multi-phase transformer. The U.S. Patent Office insisted that he submit separate applications for each invention, and he was granted seven separate patents on May 1, 1888.

Having disagreed with Edison, Tesla found a champion of AC in George Westinghouse (U.S.), who bought the rights to Tesla's patents for one million dollars plus royalties and used the technology to open, in 1896, the first large-scale AC power station. Tesla continued to devise electrical inventions as well as proposing a number of ideas that were far ahead of their time, including lighting the Earth at night by creating an artificial aurora (there is a modern proposal to reflect sunlight from satellites), transmitting pictures around the globe (television followed decades later), death rays (later experiments led to the invention of radar), and wireless communication, which became a reality in 1896.

See also: Air brake, page 88; Phonograph/Kinetoscope, page 158; Radar, page 110; Radio, page 154

Did you know?

In 1912 Tesla was offered the Nobel Prize for Physics jointly with Thomas Edison. Such was the grudge between the two men that Tesla refused the prize and it was awarded instead to physicist and inventor Nils Gustav Dalén (Sweden).

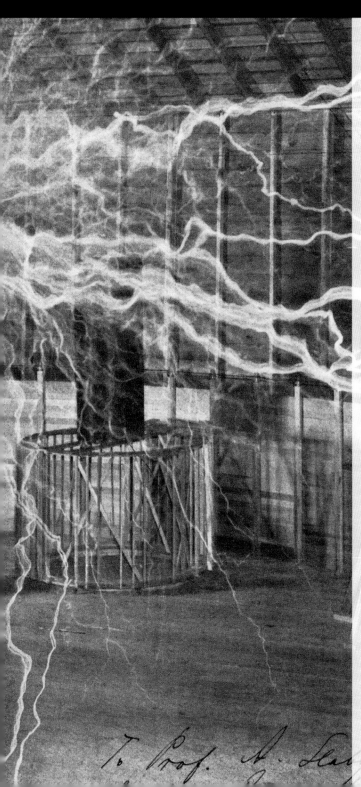

Left Generating artificial lightning in Nikola Tesla's laboratory (undated) **Above** Original Tesla induction motor (1887–88)

The inventor: Nikola Tesla

1856 Born on July 9 in Smiljan, Austria-Hungary (later Yugoslavia, now Croatia)

1875–80 Trains as an engineer at Graz Polytechnic Institute, Austria, and the University of Prague, Czechoslovakia

1881 Invents a telephone repeater while working at the Central Telegraph Office in Budapest, Hungary

1882 Conceives the principle of the rotating magnetic field, leading to the invention of the AC induction motor

1883 Makes his first AC motor

1884 Immigrates to the U.S., and works for Thomas Edison

1887 Founds the Tesla Electric Company. Files patents on October 12 for inventions including single- and multi-phase AC motors, a distribution system, and a multi-phase transformer. The U.S. Patent Office insists that he file seven separate applications, which are all granted on May 1, 1888. Tesla's patents are upheld

despite disputes over precedence, including one from Galileo Ferraris (Italy) relating to multi-phase transmission

1888 Sells rights to his patents to George Westinghouse (U.S.)

1890–95 Invents the high-frequency Tesla coil and an air-core transformer

1900 Sets up a tower in Denver to communicate with other worlds

1912 Refuses the Nobel Prize for Physics

1928 Files his last patents, for vertical takeoff and landing aircraft

1943 Dies on January 7 in New York, aged 86, having spent several years as a recluse in the New Yorker Hotel living on a pension from the Yugoslav government. He dies with no fewer than 14 honorary doctorates and 112 U.S. patents to his name. The SI unit for magnetic flux density, the tesla, is later named after him *See also: Battery, page 230*

In simple terms, a magic eye is a sensor that triggers devices such as burglar alarms, security lights, and automatic doors. The working part of a magic eye is the photoelectric cell, invented by Julius Elster and Hans Geitel c. 1893.

MAGIC EYE

The magic eye, or electric eye, performs numerous useful functions, from opening doors and triggering alarms to switching streetlights on when it gets dark and off when it gets light again. As well as these everyday functions, magic eyes, more scientifically known as photoelectric cells or photocells, are also used in control engineering, precision measuring devices, counting devices, solar panels, light meters, and scanners, as well as being used in solar batteries as sources of electric power for rockets and satellites.

The principle behind the photoelectric cell is known as photosensitivity: the electrons in some semimetallic materials become energized when exposed to light or other forms of electromagnetic radiation, creating an electric current that flows through the cell. Photo*conductive* cells work on a similar principle, except that, rather than a current being created by light, the resistance of the cell increases or decreases, causing changes in the current that is able to pass through it. Because they are so sensitive, photoelectric and photoconductive cells can be used not only in switches (where they are used to turn a device on or off if the intensity of light changes or reaches a particular level), but also in meters, where the amount of light can be measured through the amount of energy produced in the cell.

American scientist G.R. Carey is said to have invented a photoelectric cell as early as 1875, but it was not put to practical use until the 1920s, and the first practical photoelectric cell was invented by Julius Elster and Hans Geitel (both Germany) c. 1893. These two eminent physicists worked together at the Herzogliches Gymnasium in Wolfenbüttel, Germany, where they also invented a photometer and an electrical transformer, and made several important discoveries in the fields of electricity and radiation. They did not take out patents on any of their inventions because they believed that these should be of benefit to everyone.

Below Automatic doors triggered by a magic eye **Above** A solar-powered car created at the California Polytechnic Institute is put through its paces at a solar and electric energy exposition in Phoenix, Arizona (May 1992)

Did you know?

Asked if they can quote one of Einstein's equations, most people would say "E=mc². " The science behind the energy transfer in the photoelectric process is expressed in another of Einstein's equations: $h\nu = A + 1/2\, m_e v^2$.

The inventors: Julius Elster & Hans Geitel

1854 Elster is born Johann Phillipp Ludwig Julius Elster on December 24 in Bad Blankenburg, Germany

1855 Geitel is born Hans Friedrich Geitel on July 16 in Brunswick, Germany

1875–77 Elster and Geitel both study at Heidelburg University

1877–79 Elster and Geitel both study at Berlin University

1880 Geitel begins teaching at the Herzogliches Gymnasium in Wolfenbüttel, near Brunswick

1881 Elster begins teaching at the Herzogliches Gymnasium

1889 Elster and Geitel perform the first of 20 experiments into the photoelectric effect and discover that the effect can be induced by visible as well as ultraviolet light

c. 1893 Elster and Geitel invent the first photoelectric cell. They later invent a photometer and a transformer

1899 Elster and Geitel determine the electric charge on raindrops from thunderclouds. Other scientific achievements include proving that lead is not radioactive and that radioactive ionization causes atmospheric conductivity. The University of Göttingen awards Geitel an honorary Ph.D.

1915 Geitel is awarded an honorary doctorate in engineering by the University of Brunswick

1920 Elster dies on April 6 in his birth town, Bad Blankenburg, aged 65

1923 Geitel dies on August 15 in Wolfenbüttel, aged 68

Introduced to cinematography by his father on the feature film *Out of Africa*, Matthew Allwork went on to invent some of the world's most sophisticated sports photography equipment, including the "Stumps Cam" and the "Jockeycam."

STUMPS CAM

Matthew Allwork (England) was the son of an aerial cinematographer. While studying law at the University of London he took time out to join his father, Peter, on the film unit for *Out of Africa* and was so taken with motion photography that he abandoned his law studies to join the family firm, Aerial Camera Systems (ACS). Matthew immediately began to combine his camera skills with his love of sports, broadening the scope of the company from pure aerial cinematography to cutting-edge sports coverage.

Determined to come up with innovative and exciting ways of televising sport and large-scale events, Allwork invented numerous specialist cameras and rigging devices that are now the industry standard for televising everything from soccer and football to horse racing and rock concerts. He commented: "We're putting a camera where it's never been before, in order to cover sport in a more dynamic and versatile way." He was instrumental in introducing two gyrostabilized cameras to film and broadcast television, adapting them for use with helicopters, airships, and the Rail System, the latter a co-invention that has become a familiar sight when tracking runners at the Olympics and other athletic events.

Allwork revolutionized the way sports is televised, bringing viewers closer to the action than had ever been thought possible, with inventions such as the Helmet Camera, a robust miniature camera designed to be attached to a jockey's helmet and soon dubbed the "Jockeycam." Another gadget was the Mini Pan & Tilt Head, a remotely operated camera mount originally designed to be attached to the ice hockey net for the Lillehammer Winter Olympics and since used for soccer, rugby, tennis (attached to the umpire's chair), and horse racing (above the starting gate). It has also been adapted for cricket, where the Stumps Cam, made from carbon fiber and Kevlar to withstand the fastest of bowlers, provides an unprecedented view of the game from *inside* the stumps.

See also: Kevlar, page 250

The inventor: Matthew Allwork

1963 Born Matthew Julian Allwork on May 23 in Weybridge, Surrey, the son of an aerial cinematographer. He is later educated at St. George's College, Weybridge, and London University

1982 Helps adapt the Wescam gyrostabilized camera for use with helicopters (and, later, blimps, wirecam, and track) and becomes instrumental in introducing gyrostabilized cameras to film and broadcast television

1990 Becomes managing director of Aerial Camera Systems (ACS)

1992–2000 Develops, with colleagues, the Rail System for tracking runners at athletic events

1993 Invents the Mini Pan & Tilt Head for the Lillehammer Winter Olympics, later adapted for various uses including the Stumps Cam

1994 Awarded an Emmy for the Hockey Net Cam (officially the Mini Pan & Tilt Head) at the Winter Olympics in Lillehammer

1998 Invents the Mini Vertical Track. Designed to lift a small camera up to 23 ft. (7 m) off the ground,

it provides spectacular camera angles for soccer (behind the goal), pop concerts, and various other events. Invents and patents the Helmet Camera, aka Helmet Cam, aka Jockeycam

1999 Helps adapt the Gyron gyrostabilized camera for use with tracking vehicles, revolutionizing the filming of horse racing. Wins an Emmy for his Wirecam camera-operating on the Janet Jackson concert in Madison Square Garden. ACS wins the Royal Television Society Award for product innovation in televised sports (Wirecam for Channel 4 Racing at Cheltenham), and Channel 4 wins a BAFTA for its coverage of the 1998 Epsom Derby using ACS cameras

2000 Wins an Emmy for his work on the Sydney Olympics. ACS begins work on a Stumps Cam that will pan through 360 degrees

2003 Killed in a helicopter accident on March 26, at the age of 39, while filming the Dubai Endurance Horse Race in the United Arab Emirates. Posthumously awarded an Emmy for his work on the 2002 Winter Olympics in Salt Lake City

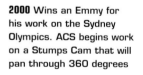

Did you know?

As well as in sports, Allwork's camerawork has been seen in numerous cutting-edge television and movie commercials, and in feature films including *Indiana Jones and The Last Crusade*, *Mary Shelley's Frankenstein*, *A View to a Kill*, *Robin Hood: Prince of Thieves*, and *The Fourth Protocol*.

Speaking at Allwork's funeral in 2003, Andrew Franklin (Britain), executive producer of Channel 4 racing, praised Allwork's Jockeycam and said: "If he'd have been let loose on *Songs of Praise* it wouldn't have been long before Matthew would have come up with the 'Priestcam.'"

Opposite top Matthew Allwork (1963-2003), inventor of numerous cutting-edge sports photography systems **Opposite below** The Stumps Cam **Below left** The Rail System in use at the Sydney Olympics (2000) **Below right** U.S. sprinter Maurice Green at the Sydney Olympics (2000)

Serra is a unique light-bending system invented by Peter Milner in 1989. It is used in the automotive industry and in architecture, including for high-tech mirrors and as an "intelligent" window glass, reducing glare and increasing illumination.

SERRA

The invention of the Serra light-bending system arose from an earlier technology, the Fresnel prism, invented in the early 19th century by Augustin Jean Fresnel (France). Nearly 200 years later, Peter Milner (England) was experimenting with the use of Fresnel prisms to create a revolutionary mirror for cars, when his research led to the invention of a new technology that he marketed under the brand name Serra.

Serra consists of what Milner calls "an array of micro-replicated prismatic structures": in effect, a combination of Fresnel prisms precisely mounted in such a way that they bend light in the desired direction by reflection and/or refraction. The initial application was the Serraview prismatic mirror, which could be mounted within a vehicle, giving huge aerodynamic advantages over conventional exterior wing mirrors. Other vehicular applications are Serrascope, a prismatic refracting periscope that gives drivers a better view of the exterior of the vehicle for low-speed maneuvers, and Serravista, which enables drivers to "see around" the windshield pillars.

Milner was encouraged by an architect to turn his attention to buildings, and in 1993 invented Serraglaze, an "intelligent" window glass. The level of daylight entering a room through a conventional window falls off rapidly within a few feet of the glass, but not so with Serraglaze. Milner explains: "Outside light arriving at a Serraglaze window more or less horizontally (which is the light that penetrates deepest into a room and also provides the occupants with a view out) passes through undisturbed, thereby preserving these benefits. But light arriving from higher elevations, which is normally largely wasted by falling on the floor just inside the window, and which often causes glare for those near the window, is intercepted and redirected deep inside the room where light is most needed."

The inventor: Peter Milner

1941 Born on November 3 in Evesham, England

1953–64 Attends Prince Henry's Grammar School, Evesham, and University of Sheffield

1964–83 Works as a car design engineer with Rootes, Chrysler, Volkswagen, Peugeot, and DeLorean

1989 Invents the Serraview prismatic mirror, the first application of his Serra technology, and files his patent on December 16 (granted June 29, 1994)

1992 Awarded a British government technology award for Serra

Left A bust of Italian physicist Macedonio Melloni gazes through a Fresnel lens **Above** Detail of a Fresnel lens, invented by French physicist Augustin Jean Fresnel **Below** A modern car sporting Peter Milner's Serra interior mirrors

1993 Invents Serraglaze and files his patent on May 3, 1994 (granted November 14, 2001)

1999 Invents Serrascope and files his patent on October 9, 2000 (patent pending). Approaches inventors' agency Inventorlink to assist the location of financial backing for Serraglaze

2001 Establishes Bending Light Ltd. to develop and market Serra

2002 Invents Serravista, and files his patent on January 10, 2003 (patent pending). Serraglaze goes into production in Apolda, German

2003 Dies in December after a brave battle with cancer

Bar codes are now so commonplace that they often pass unnoticed, but the familiar cluster of thick and thin stripes might have looked very different—the inventors originally envisaged bar codes as a series of concentric circles forming a bullseye.

BAR CODE

In the late 1940s a postgraduate student named Bernard Silver (U.S.) heard that the Drexel Institute of Technology, where he was studying, had turned down a request from the president of a chain of food shops to develop a means of automatically collecting product information at the checkout. The institute may not have been interested in the idea, but Silver was; he told fellow student Norman Woodland (U.S.) and together they began to research the concept.

Their first idea was to use patterns of fluorescent ink that would glow under ultraviolet light, but this proved to be impractical and expensive. Then Woodland came up with the idea of a label based on the principle of Morse code, except that instead of dots and dashes he proposed thick and thin lines that could be read by a scanner. This embryonic idea was very close to the modern bar code, but Woodland and Silver thought that it would be difficult to scan and they developed the idea further; in 1949 they filed a patent for a data code in the form of concentric circles, which meant that the scanner did not have to be held parallel to the bar code. (Modern laser scanners overcome the problem by scanning in several directions at once.)

They then built a prototype scanner, which, although it scorched the codes it was reading, proved that the idea worked. By this time Woodland was working for IBM, which twice offered to buy the patent rights—eventually Philco bought the rights in 1962, later selling them to RCA. However, IBM had the last laugh by being the first to market a practical code and scanning system. Still working for IBM in the 1970s, Woodland, together with George Laurer (U.S.), developed a sophisticated 12-digit code now known as the Universal Product Code, which was approved in 1973. The following year a packet of chewing gum became the first item to be sold using a bar code, when processed at the Marsh Supermarket in Troy, Ohio, at 8:01 a.m. on June 26, 1974.

Did you know?
European bar codes have one more digit than American ones. America uses the 12-digit Universal Product Code, while Europe uses the 13-digit European Article Number.

The name IBM (International Business Machines) inspired the name of the computer HAL in the film *2001: A Space Odyssey.* Stepping back one letter in the alphabet from each of the initials IBM produces the name HAL.

The inventors: Bernard Silver & Norman Woodland

1924 Bernard Silver born

1921 Norman Woodland born Norman Joseph Woodland

1949 Woodland and Silver file a patent for: "Classifying apparatus and method" (granted October 7, 1952)

1962 Philco buys the rights to Woodland and Silver's patent. Silver dies, aged 38

1973 Woodland helps to design the Universal Product Code for IBM. IBM produces the first practical bar code scanner

1974 The first item to be sold using a bar code is processed on June 26 in U.S.

1992 Woodland is awarded America's National Medal of Technology

Opposite top Bar code reader in use at a supermarket checkout in the U.S. **Below** A bar code being laser scanned

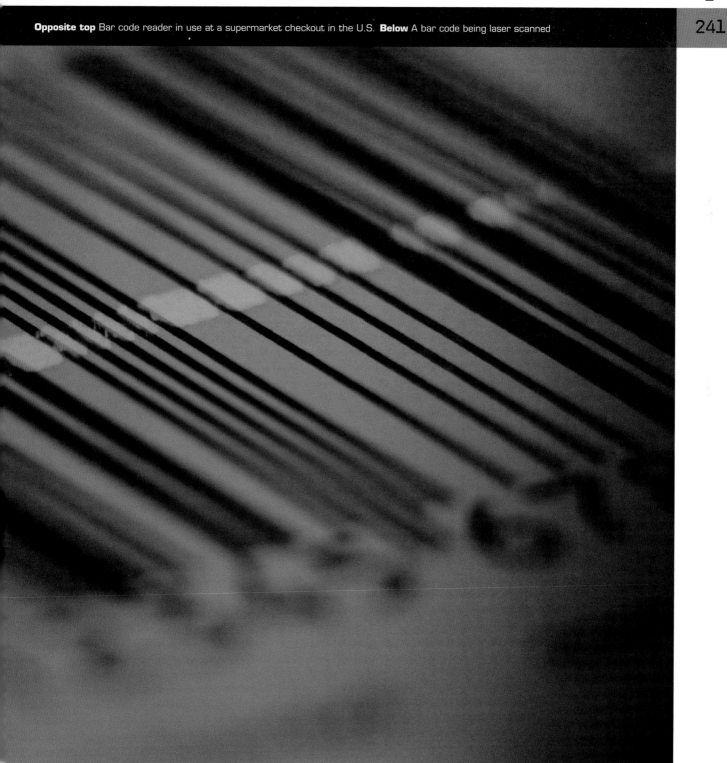

Familiar to most people from Cold War-era science fiction films and television programs, the crackling Geiger counter is more correctly known as a Geiger-Müller counter, named after its inventors, Hans Geiger and Walther Müller.

GEIGER COUNTER

Geiger counters have been used in countless fictional situations to assess atmospheric radioactivity on alien planets and spaceships or after nuclear attacks and accidents. The counter would be quite capable of measuring radiation in such situations, but that was not the reason it was invented—it has a far more useful function as a precision laboratory instrument for detecting and measuring various particles, including alpha particles, electrons, and, of course, ionizing photons, better known to sci-fi fans as nuclear radiation.

In 1906, immediately after gaining a Ph.D. in physics, Hans Wilhelm Geiger (Germany) traveled to England to work at Manchester University under Ernest Rutherford (New Zealand), who was to become known as the father of nuclear physics. Rutherford had already established that radioactive elements gave off two types of radiation, one negatively charged and the other positively charged, which he called alpha and beta particles. Together, Rutherford and Geiger invented an instrument that could detect and count alpha particles, and which became a vital tool in Rutherford's pioneering work on determining the composition of the atom. This first counter comprised a glass bulb within which was a small tube containing a wire connecting two high-voltage electrodes. When a gas was introduced to the bulb the current caused it to ionize, creating pulses of electrical current—the sound of these "sparks" was amplified by a loudspeaker to produce the clicking sounds familiar to moviegoers.

In 1912 Geiger returned to Germany to take up a post as head of the Physikalisch Technische Reichsanstalt in Berlin. He later became a professor at Kiel University, where, in 1928, he and Walther Müller (Germany) significantly improved the original particle counter to detect and count a wider range of radioactive particles with greater accuracy—a direct descendant of the Geiger-Rutherford machine, this was the first of the modern Geiger-Müller counters.

The inventor: Hans Geiger

1882 Born on September 30 in Neustadt-an-der-Haardt, Germany (now Neustadt-an-der-Weinstrasse), the son of a professor of philology

1906 Receives a Ph.D. in physics from the University of Erlangen, Germany

1906–12 Works under Ernest Rutherford (New Zealand) at Manchester University, England

1908 With Rutherford, invents the first electrical alpha particle counter (in order to compare the results of an

Did you know?

In a patent granted in 1968, Arthur Pedrick (England) suggested the use of a Geiger counter to find lost golf balls. His patent related to: "Improvements in the flight direction and location of golf balls," and described a golf ball containing two spheres of metal or wire mesh that would reflect "high frequency radiation emitted from a maser, magnetron or klystron device carried by the golfer." The reflected radiation would be detected by "a Geiger counter or other radioactivity-sensitive device, which will thus indicate the approximate position of a golf ball so constructed for "'semi-active homing.'" *See also: Arthur Pedrick, page 272; Golfing inventions, page 168*

Opposite top A Geiger counter, used for detecting radioactivity by its ionizing effect on a gas at low pressure (contained within the cylindrical tube) **Below** Testing for radiation leaking to the surface after underground atomic tests, in the film *Blue Sky* (1990)

experiment with those of Erich Regener, Germany, who used a different type of counter)

1912 With Rutherford and John Michael Nuttall (England) demonstrates the mathematical relationship between alpha particles and the radioactive nucleus

emitting the particles, now known as the Geiger-Nuttall rule (revised by Geiger in 1921). Appointed head of the Physikalisch Technische Reichsanstalt in Berlin

1925 Appointed professor at Kiel University

1928 With Müller, invents an improved particle counter known as the Geiger-Müller counter

1929 Awarded the Hughes Medal of The Royal Society for "the invention and development of methods of counting alpha and beta particles"

1938 Awarded the Duddell Medal of the Physical Society, London, for contributions to scientific instrumentation

1945 Dies on September 24 in Potsdam, Germany, aged 62

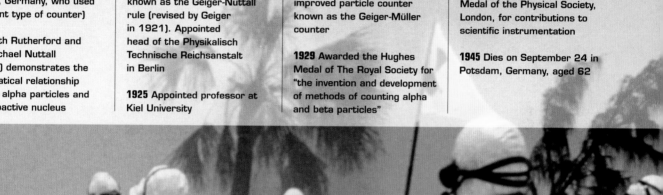

The invention of the transistor in 1947 represented a huge leap forward in electronics and communications technology, and was considered such an important development that its inventors shared the 1956 Nobel Prize for Physics.

TRANSISTOR

The invention of the transistor was of vital importance not only for the birth of the transistor radio, but also for the development of all electronic equipment, from mobile phones and computers to robots and satellites. Prior to the Second World War, William Shockley and Walter Brattain (both U.S.) had been conducting research at Bell Telephone Laboratories into the use of semiconductors to replace the bulky, inefficient thermionic valves that were then standard in electronic equipment. Their work was interrupted by the war, but afterward they were joined by John Bardeen (U.S.) and carried on with their research.

Shockley described his team's method as "creative failure" and said that they were "using failures as opportunities to learn and move ahead." One such creative failure was Shockley's attempt to create a field-effect amplifier. Bardeen modified the experiment to introduce a controlling current from a third contact, leading to the invention of the point-contact transistor, which he and Brattain patented as a "three-electrode circuit element utilizing semiconductive materials." Shockley was not named on this patent, but received a separate patent, in his own name, for the invention of the junction transistor—an improvement on the point-contact transistor that involved replacing the point contacts with rectifying junctions. (Junction transistors remained the industry standard until transistors were superseded by integrated circuits, or microchips, in the late 1960s.)

Shockley had told Bardeen and Brattain that as the originator of the idea he should be named as the inventor of the original point-contact transistor, to which Brattain reportedly replied: "There's glory enough in this for everyone." He was right, because not only was Shockley named as the inventor of the improved transistor, but all three shared the 1956 Nobel Prize for Physics in recognition of these inventions and their subsequent work on transistors.

See also: Microchip, page 246; Radio, page 154

transistors **PHILIPS**

c'est plus sûr !

Above Poster by Pierre Fix-Masseau advertising Philips transistor radios (1950s) **Below** John Bardeen, William Shockley, and Walter Brattain at Bell Telephone Laboratories (1948)

The inventors: William Shockley, John Bardeen, & Walter Brattain

1902 Walter Brattain is born on February 10 of American parents in Amoy, China, and later studies at the University of Oregon

1908 John Bardeen is born on May 23 in Madison, Wisconsin, and later studies at the University of Wisconsin

1910 William Shockley is born on February 13 of American parents in London, England, and later studies at the California Institute of Technology

1929 Brattain gains a Ph.D. from the University of Minnesota and joins Bell Telephone Laboratories as a research scientist later that year

1936 Bardeen graduates from Princeton with a Ph.D. in mathematical physics. Shockley graduates from Massachusetts Institute of Technology with a Ph.D. in physics and joins Bell Telephone Laboratories

1936–42 Shockley and Brattain begin research into semiconductors

1938 Bardeen is appointed assistant professor of physics at the University of Minnesota

1939–45 During the war Bardeen works as a physicist for the Naval Ordnance Laboratory. Brattain and Shockley are assigned to Columbia University to research the magnetic detection of submarines

1945 Bardeen becomes a research physicist at Bell Telephone Laboratories, where he joins Brattain and Shockley

1947 Brattain and Bardeen invent the point-contact transistor (patent filed 1948, granted 1950). Shockley conceives the junction transistor

1950 Shockley tests and patents the junction transistor

1956 Shockley, Brattain, and Bardeen share the Nobel Prize for Physics

1972 Bardeen shares the Nobel Prize for Physics with Leon Cooper and John Schrieffer (both U.S.) for their work in low-temperature superconductivity

1987 Brattain dies on October 13 in Seattle, Washington, aged 85

1989 Shockley dies on August 12, aged 79, at Stanford University, where he had been a professor from 1963 to 1975

1991 Bardeen dies of a heart attack on January 30 in Boston, Massachusetts, aged 82

Valve computers had relatively little computing power and frequently burned out, and each computer filled an entire room. When transistors replaced valves, computers became smaller and more reliable. Then Jack Kilby invented the microchip...

MICROCHIP

The microchip, like that other hugely significant invention the telephone, was being developed in two places simultaneously and, as with the telephone, being first to the patent office proved to be vitally important. Having trained as an electrical engineer and gained experience in miniaturization by working in the established field of radio and television parts manufacture, Jack Kilby (U.S.) moved to Texas Instruments (U.S.) in May 1958. His first job was to research ways of miniaturizing electronic components, and it was his unconventional thinking that brought about a complete revolution in computer engineering.

Standard practice was to make all the components of a circuit separately and then join them together as required. Instead of trying to miniaturize each individual component, Kilby began to wonder what would happen if they were all made from the same material, and then conceived the idea of using a semiconductor material to do so. (A semiconductor, such as silicon, can be made to behave either as a conductor or as an insulator, which means that it can perform various electronic functions.) Kilby used slivers of silicon to create a circuit in which all the components were integrated, and in August 1958 he demonstrated his invention to Texas Instruments. Kilby filed several patents for this integrated circuit, or microchip, on February 6, 1959.

Meanwhile, Robert Noyce (U.S.), and fellow scientists at Noyce's company Fairchild Semiconductor, had been developing a more robust and more easily mass-produced microchip known as a planar integrated circuit, for which they filed a patent on July 30, 1959. The Fairchild chip was the first to be produced commercially, for which reason Noyce is often named as co-inventor of the microchip, but Kilby was the first to patent his invention. After a lengthy legal battle, the U.S. courts ruled that, although Fairchild had the rights to certain techniques, Jack Kilby's patent gave him precedence as the inventor of the microchip.

See also: Telephone, page 140

Above Jack Kilby, inventor of the integrated circuit, surrounded by electronic devices at the Texas Instruments headquarters, Texas (October 1982) **Below** An ant with a microchip (2001)

The inventors: Jack Kilby & Robert Noyce

1923 Kilby is born Jack St. Clair Kilby on November 8 in Jefferson City, Missouri, the son of an electrical engineer. He later graduates in electrical engineering from the University of Illinois and then receives a master's degree from the University of Wisconsin

1927 Noyce is born Robert Norton Noyce on December 12 in Burlington, Iowa, and later studies at the Massachusetts Institute of Technology (MIT)

1948 Kilby joins Centralab, a radio and television parts manufacturer in Milwaukee

1957 Noyce co-founds Fairchild Semiconductor

1958 Kilby joins Texas Instruments (TI, U.S.) in May. Invents the first monolithic integrated circuit, or microchip, and in August demonstrates a prototype to his superiors at TI

1959 Kilby files patents on February 6 for "miniaturized electronic circuits" (granted 1964). Noyce and colleagues file a patent on July 30 for a planar integrated circuit

1961 Fairchild Semiconductor produces the first commercially available microchips

1967 With Jerry Merryman and James van Tassel (both U.S.), also of Texas Instruments, Kilby invents the first handheld electronic calculator, for which they file a patent on December 21, 1972 (granted June 25, 1974), having abandoned two previous applications *See also: Calculator, page 126*

1970 Kilby leaves TI to become a freelance inventor and consultant. By the turn of the century he has 60 patents to his name, mainly relating to integrated circuits

1978 Kilby becomes Distinguished Professor of Electrical Engineering at Texas Agricultural and Mechanical University

1990 Noyce dies on June 3 in Austin, Texas, aged 62, having co-founded another successful microchip manufacturing company, Intel

2000 Kilby is awarded the Nobel Prize for Physics

By the middle of the 20th century barbed wire was seen as a symbol of oppression, but when Joseph Glidden invented the modern form in 1873 its effect was to give farmers their defense against the cattlemen of America's Wild West.

The inventor: Joseph Glidden

1813 Born Joseph Farwell Glidden on January 18 in Charlestown, New Hampshire. Later attends Middlebury Academy, Vermont, and Lima Seminary, New York, and then works as a teacher and on his father's farm, before buying a 600-acre (240 ha) farm of his own in De Kalb County, Illinois

1873 Sees Henry M. Rose's (U.S.) patent barbed wire at a country fair. Invents an improvement and files a patent for: "Wire Fences" on October 27 (granted November 24, 1874)

BARBED WIRE

Farmer Joseph Glidden (U.S.) was by no means the first person to invent and patent barbed wire, but he was the first to make a success of it. Various forms of wire fencing were patented in the early 19th century, some of the later ones incorporating a twist that acted as a primitive (and blunt) form of barb. The first wire fencing to incorporate a separate barb was invented by Lucien B. Smith (U.S.), whose patent (granted June 25, 1867) describes barbs protruding from blocks of wood fixed at intervals along the wire, but there is no evidence that he ever manufactured his invention. A month later, on July 23, William Donison Hunt (U.S.) was granted a patent for another type of barbed wire (sometimes cited as the first), and in February 1868 Michael Kelly (U.S.) was granted a patent for a wire known as Kelly's Diamond because of its diamond-shaped barbs.

Various other types followed, including a barbed fence patented by Henry M. Rose (U.S.) in May 1873. That summer, Glidden and two friends, Isaac Ellwood and Joseph Haish (both U.S.), saw Rose's fence exhibited at a country fair. Glidden and Haish thought they could improve on Rose's invention, and both filed patents for their improvements—Glidden in October and Haish in December. Glidden's version involved threading the barbs directly onto a two-strand twisted wire, and is the first example of the modern form of barbed wire. Haish challenged Glidden in court, lost, and went on to invent "S" barbed wire (patented in 1875), which made Haish a fortune, although it was never a serious competitor to Glidden's wire.

Barbed wire quickly spread across the plains of the American West as farmers fenced their properties, much to the annoyance of the cattle barons who had previously driven their herds wherever they chose, making arable farming all but impossible. Fence-cutting wars and murders ensued, but to no avail: Barbed wire had opened up the West to settlers, and the wire, like the settlers, was there to stay.

Did you know?

There are more than 1,500 different types of barbed wire, making it an ideal subject for collectors—in America, an 18-inch piece of one rare variety of barbed wire fetched $65 at auction.

So many types of barbed wire have been patented that there is an entire book devoted to the subject: *Early United States Barbed Wire Patents* (1966) by the appropriately named Jesse James.

1874 The U.S. Patent Office declares in favor of Glidden after a challenge from Joseph Haish (U.S.)

1875 Glidden sells a half interest in the patent rights to Isaac Ellwood (U.S.) and forms partnership Barb Fence Co.

1876 Sells out to Charles Washburn (U.S.) of the Washburn & Moen Manufacturing Co. for a fee of $600 plus royalties

1891 After several lawsuits, the U.S. Supreme Court upholds the validity of Glidden's

patent, which, ironically, expires that same year

1906 Glidden dies on October 9 in De Kalb, aged 93

Opposite Inventor Joseph Glidden features in an advertisement for Glidden Steel Barb Fence Wire (c. 1877)
Below A cowboy carefully adjusts a barbed-wire fence as his horse looks on near Ketchum, Idaho (c. 1986)

249

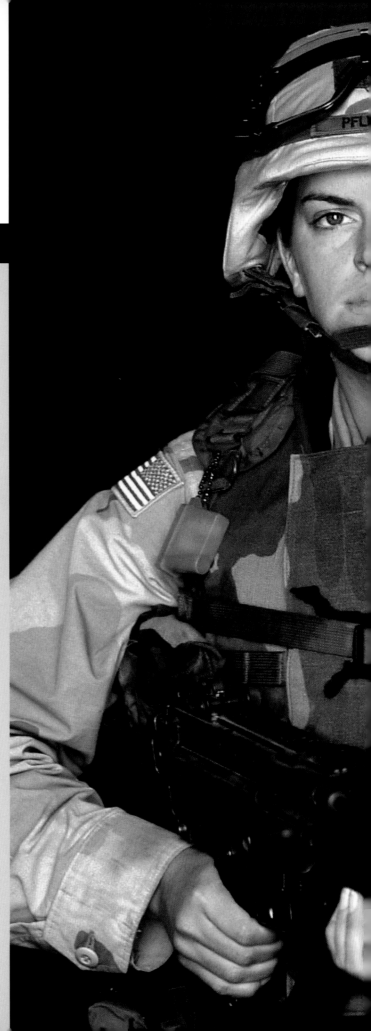

As a female scientist working for a large corporation in the 1950s, Stephanie Kwolek encountered cynicism and mysogyny from many of her male colleagues—until she invented Kevlar, a synthetic fiber with almost miraculous properties.

KEVLAR

Kevlar sounds simply too good to be true. Invented in 1963 by scientist Stephanie Kwolek (U.S.) as part of her research into synthetic textile fibers for chemical company DuPont (U.S.), it is an extremely strong, lightweight, fire-resistant fiber that has found its way into everything from bulletproof vests and bridge cables to racing cars, lifeboats, and fishing lines.

Kwolek, who already had a degree in chemistry, began working for DuPont part time in order to fund her way through medical school. But she found her work at DuPont so interesting that instead of becoming a doctor she decided to stay on, joining the company's Pioneering Research Laboratory full time in 1950 and researching the manufacture of very strong fibers from long chains of molecules. Kevlar originated not as the planned result of an experiment, but because Kwolek was inquiring enough to investigate the unexpected. She was researching an improved means of joining certain simple molecules (monomers) into longer chains (polymers) by fusing them at low temperatures, but the result was an opaque solution, not the clear liquid she was expecting. Instead of ditching the solution and starting again, Kwolek tried spinning it into fibers and found that she had created an entirely new type of synthetic material, described in her patent as "crystalline, linear condensation polyamides."

Analysis showed that the fiber she had created was five times as strong as steel of the same weight, and that when woven into a textile or fabricated into a composite material, it could absorb energy and prevent cracks from spreading from the point of impact. The process has been described as being like that of a spider's web, in that the fibers at the point of impact stretch rather than break, dissipating energy by causing other fibers beyond the impact to stretch as well—properties so remarkable that Kwolek has been described as a modern alchemist for conjuring up Kevlar.

Did you know?

Kwolek and Morgan's patent covered an entire class of crystalline polyamides—the one that became Kevlar is described as: "A linear polyamide selected from the group consisting of polymetaphenylene adipamide, polymetaphenylene suberamide, and polymetaphenylene sebacamide, said polyamide having a viscosity of at least 0.8 and a crystallinity index above 50."

Left A female gunner wears Kevlar helmet and body armor in Afghanistan (February 2002)
Above Kevlar yarn bobbins at the DuPont Kevlar plant, Virginia

The inventor: Stephanie Louise Kwolek

1923 Born on July 31 in New Kensington, Pennsylvania

1942–46 Studies at the Carnegie Institute of Technology and graduates with a degree in chemistry. Later begins studying for a degree in medicine, working for DuPont to fund her medical studies

1950 Joins DuPont's Pioneering Research Laboratory full time

1963 Invents Kevlar and, together with colleague Paul Winthrop Morgan (U.S.), files a patent on April 25 for: "Process for the production of a highly orientable, crystallizable, filament-forming polyamide" (granted November 22, 1966). The patent states: "This invention relates to a

novel class of crystalline, linear condensation polyamides and to a manufacturing process therefor"

1971 Kevlar is marketed commercially

1974 Kevlar is officially registered as a U.S. trademark

1986 Kwolek retires with 27 patents to her name but continues to work part time for DuPont, championing the cause of women scientists

1997 Awarded the Perkin Medal of the American Chemical Society for outstanding achievements in applied chemistry, becoming the second woman to receive the award

Polythene is now such a common material that it is hard to imagine the importance of its invention in 1933. Since playing a vital role in the wartime development of radar, it has revolutionized consumer life in countless plastic products.

POLYTHENE

Polythene, more correctly known as polyethylene, was first produced by Eric Fawcett and Reginald Gibson (both Britain), two chemists working for ICI (Imperial Chemical Industries, England). Fawcett and Gibson were attempting to form a polymer of ethylene, which many scientists deemed impossible, by using high temperatures and pressures, and on Friday, March 24, 1933, they set up a reaction, which they left over the weekend. The following Monday Fawcett wrote in his notebook: "Waxy solid found in reaction tube"—the first record of polythene. They were unable to repeat their success and were diverted by ICI into other research, but Fawcett realized that it was a historic discovery and announced it at a conference in 1935; fortunately no one believed him, because polythene had not yet been patented.

Then Michael Perrin, John Patton, and Edmond Williams (all Britain) resumed work on high-pressure polymerization, and on December 20, 1935, after a number of failures, they succeeded in replicating the original experiment. They discovered that the key to success had been the accidental leakage of oxygen into the pressure chamber, which must also have occurred in the 1933 experiment. Once this was established they were able to produce polythene at will in significant quantities, and patented their method in 1937 in the names of all five chemists.

Fawcett's claim to have achieved the supposedly impossible was vindicated, and production of polythene began on a small scale in 1937, with a full-scale plant opening on September 1, 1939, the day Germany invaded Poland and started the Second World War. One of the first uses for polythene was as an electrical insulator in wartime radar, and it was later used to insulate the first round-the-world telephone cable. The first consumer items to be made of polythene were washing-up bowls in 1948, followed by products ranging from bottles, buckets, and barrows to supermarket packaging and the humble polythene bag.

See also: Radar, page 110; Tupperware, page 68

ICI: key inventions

1926 ICI is formed according to the "Aquitania Agreement." The agreement is drawn up on a six-day voyage from New York to Southampton aboard the Cunard liner *Aquitania* and proposes the merger of Brunner Mond & Co., Nobel Industries, United Alkali (not represented on the *Aquitania*), and British Dyestuffs Corporation as Imperial Chemical Industries. The new company is officially incorporated on December 7

1932 John Crawford (Britain) invents a process of synthesizing plastics that leads to the industrial manufacture of polymerized methyl methacrylate, better known by its tradename Perspex (from the Latin for "to see through"). A patent for manufacturing Perspex is granted to Rowland Hill of ICI in 1933, and one of its first uses is in the cockpit canopies of Spitfire fighter planes

1933 Eric Fawcett & Reginald Gibson (both Britain) produce the first polythene (polyethylene)

1937 Polythene is patented under the title: "Improvements in or relating to the Polymerisation of Ethylene" by ICI scientists Eric William Fawcett, Reginald Oswald Gibson, Michael Wilcox Perrin, John Greves Patton, and Edmond George Williams (all Britain) (provisional patent filed February 4, 1936, complete specification filed February 4, 1937, granted September 6, 1937)

1941 John Rex Whinfield and James Tennant Dickson of the Calico Printers Association (all Britain) invent a method of producing the first polyester fiber, which is subsequently developed and produced by ICI around the world as Terylene, except in the U.S., where it is produced under license by DuPont as Dacron (patent filed 1941, granted 1946)

1959 A patent is granted for Paraquat, a weed killer invented by a team of four ICI scientists that acts on vegetable matter but is inactivated on contact with soil

1964 James Black invents the first clinically practicable beta-blocker—propranolol—as a result of researching a treatment for angina for ICI Pharmaceuticals *See also: Beta-blocker, page 194*

1967–71 Rank Hovis McDougall (RHM, England) and ICI develop the foodstuff mycoprotein, eventually manufactured by ICI under exclusive license from RHM and marketed as Quorn *See also: Quorn, page 80*

1981 ICI scientist Derek Birchall (Britain) is granted a patent for "macro-defect-free" (MDF) cement, an extremely strong, flexible cement developed after examination of the properties of molluscs and cuttlefish bone

Left A mundane but soothing use of polyethylene: a plastic bowl to soak aching feet **Above** Apparatus used to discover polythene, 1933, presented by ICI to the Science Museum in London on the 50th anniversary of the first experiment

Did you know?

Peter Allen (England), who coordinated the early development of polythene for ICI from 1936, remembered later: "We were just experimenting, and then war broke out. There was an immediate requirement for anything that could illuminate in the blackout. We found that polythene was a good carrier for luminous powders subjected to ultraviolet light, and one of the things we were doing in the first weeks of the war was devising strips of [luminous] polythene to put round policemen's helmets....Then radar took hold and we were slaving like mad to make more and more."

How different things might have been if ICI's trademark had caught on instead of the generic name polythene—in April 1936 ICI registered the trade name Alketh, which was later changed to Alkathene.

The wavy lines of the ICI logo survive from the original logo of Nobel Industries, for whom Alfred Nobel invented dynamite and gelignite. The Nobel family firm was one of the four companies that merged to form Imperial Chemical Industries. *See also: Dynamite, page 266*

Mankind's pathological desire to kill has led to increasingly inventive means of exterminating people and animals, a process that became even more efficient with the invention of weapons such as revolvers, repeating rifles, and machine guns.

GUNS

Revolver

The first commercially successful revolver was invented by Samuel Colt (U.S.) in 1835. There had been other guns with rotating magazines, including one invented by Elisha Collier (U.S.) in 1831, but none of them was a practical or commercial success. In 1830 Colt set sail for India as a merchant seaman, and while in Calcutta he came across a Collier revolver. A childhood fascination for guns was revived and he determined to improve Collier's gun. On his return to the U.S. he employed various machinists and gunsmiths to perfect his invention, and eventually made a fortune from manufacturing guns. He also went on to invent underwater mines and submarine telegraph cables.

Repeating rifle

The first practicable repeating rifle was invented by Benjamin Tyler Henry (U.S.), superintendent of the Volcanic Repeating Arms Co. In 1856 Oliver Winchester (U.S.) became principal shareholder in the company and asked Henry to improve the company's earlier Hunt-Jennings carbine. The resulting new rifle was patented by Henry on Winchester's behalf in 1860, and in 1866 the company became the Winchester Repeating Arms Co.

Machine gun

The machine gun was invented by James Puckle (England), patented in 1718 and manufactured in 1721. Various others followed, but the first commercially successful example was invented in 1861 (patented 1862) by Richard Jordan Gatling (U.S.), whose previous inventions had all been agricultural (they included a rice planter, a hemp-breaking machine, a steam plow, and later a motor plow). The Gatling Gun was a hand-cranked weapon, leaving the invention of the *automatic* machine gun, in 1883, to Hiram Stevens Maxim (U.S.–Britain), whose first patent was for a hair-curling iron.

See also: Refrigerator, page 25; Maxim gun, page 256

Above Display of Samuel Colt's pistols at a mid-19th century public exhibition **Opposite** A sculpture outside the United Nations headquarters in New York City embodies the thinking of many activists for peace

The inventors: Samuel Colt, Oliver Fisher Winchester, & Richard Jordan Gatling

1810 Oliver Winchester is born on November 30 in Boston, Massachusetts, and later runs a company making shirts in Baltimore, Maryland

1814 Samuel Colt is born on July 19 in Hartford, Connecticut

1818 Richard Jordan Gatling is born on September 12 in

Maney's Neck, North Carolina. He later founds a successful agricultural equipment business and studies medicine, but never practices

1828 Colt attends Amherst Academy, Massachusetts, but is forced to leave in 1830 after burning down part of the school during a pyrotechnic display

Did you know?

Rifle manufacturer Oliver Winchester's first business was making shirts, and his first invention, patented in 1848, was a method of cutting shirt patterns. His daughter-in-law later inherited his fortune and was told by a medium that she must continually enlarge her house to avoid the curses of those killed by Winchester rifles—the house in California was extended to at least 160 rooms by the time she died in 1922 and is now open to the public.

One gunsmith's inventive talents were put to more peaceful use in 1902 when Frank Clarke (England) invented the first automated tea-maker. *See also: Wake-up devices, page 30*

1832–35 Colt becomes a traveling entertainer under the name of Dr. Coult, demonstrating the effects of laughing gas (some accounts say "lecturing in chemistry"). The profits from this venture enable him to perfect his revolver

1835 Colt invents the first successful revolver, which he patents as a "revolving gun" (U.K. patent granted 1835, U.S. patent granted 1836)

1856 Winchester becomes principal shareholder in the Volcanic Repeating Arms Co.

1860 The first successful repeating rifle is patented by Benjamin Tyler on behalf of Winchester

1861 Gatling invents the first commercially successful machine gun

1862 Colt dies on January 10 in his birth town, Hartford, Connecticut, aged 47. Gatling is granted a U.S. patent on November 4 for: "Improvements in firearms" (U.K. patent granted 1865)

1866 Winchester produces the Winchester 66 (designed by Nelson King, U.S.) and changes the name of his company to Winchester Repeating Arms Co.

1873 Winchester produces the Winchester 73 repeating rifle, which later becomes known as "the gun that won the West"

1880 Winchester dies on December 11 in New Haven, Connecticut, aged 70

1903 Gatling dies of a viral disease akin to influenza on February 26 in New York, aged 84

The first automatic machine gun was named after its inventor, Hiram Stevens Maxim, who also patented scores of other inventions, some with more peaceful purposes than his gun—such as hair curlers and an inhaler to relieve bronchitis.

The inventor: Hiram Stevens Maxim

1840 Born on February 5 in Sangerville, Maine, the son of a farmer, woodturner, millwright, and inventor

1866–70 Works for Oliver P. Drake (U.S.), a manufacturer of scientific instruments, during which time he invents

and patents a hair-curling iron, an automatic mousetrap, and an automatic sprinkler system

1870–77 Works for the Novelty Iron Works and Shipbuilding Co., New York, during which time he invents and patents a locomotive headlight, a gas

MAXIM GUN

One of a rare breed of inventors to have emigrated from the U.S. to Britain, Hiram Stevens Maxim was born in Sangerville, Maine. His father, Isaac, was a part-time inventor and developed plans for an automatic gun and a flying machine, two ideas that Hiram would return to in later life. Hiram had little formal education and from the age of 14 worked as a coachbuilder, cabinetmaker, and mechanic before finding regular employment at his uncle's engineering works in Fitchburg, Massachusetts.

From 1866 to 1870 Maxim worked for Oliver P. Drake (U.S.), a manufacturer of scientific instruments, during which time he patented his first invention, a hair-curling iron. Then, from 1870 to 1877 he worked for the Novelty Iron Works and Shipbuilding Co. in New York, where he invented a gas engine and apparatus for gas and electric lighting. From there he moved to the United States Electric Co. (USEC), where he became chief engineer and for whom he invented a method of "flashing" electric filaments in a hydrocarbon atmosphere to ensure a uniform deposit of carbon on the filament.

In 1881 Maxim traveled to the Paris Exposition on behalf of USEC, and it was there that he revived his father's idea of inventing an automatic machine gun, after being told at the Exposition: "If you want to make a lot of money, invent something that will enable these Europeans to cut each other's throats with greater facility." He moved to England and set up a workshop in Hatton Garden, London, where in 1883 he perfected his invention, the Maxim Gun: a belt-fed single-barreled gun capable of more than 600 rounds per minute, using the recoil of each shot to load the next cartridge. In 1884 he formed the Maxim Gun Company, which eventually became part of Vickers Ltd. Four years later Maxim patented another martial invention, the smokeless gun cartridge (which meant that gunners could actually see their targets), developed between 1885 and 1888 at the behest of Lord Wolseley, commander in chief of the British Army.

See also: Refrigerator, page 25; Guns, page 254

engine, and a machine for generating gas to be used for lighting

1878 Begins working for the United States Electric Co., for whom he invents a method of carbonating light bulb filaments, which he fails to patent

1883 Invents the Maxim Gun: the first automatic machine gun. Granted a GB patent for: "Improvements in machine or battery guns and in cartridges for the same and other firearms" (U.S. patent granted 1885)

1888 Invents and patents

smokeless cartridges (U.S. patent 1890)

1884 Forms the Maxim Gun Company, which eventually becomes part of Vickers Ltd.

1889–94 Invents and develops a steam-powered aircraft, which is not successful

1900 Naturalized a British subject

1901 Receives a knighthood

1916 Dies on November 24, 1916, in Streatham, London, aged 76, with more than 270 British and U.S. patents to his name

Below left Sir Hiram Maxim with the machine gun that bore his name (undated) **Below right** The inventor as pugilist (c. 1900)

Did you know?

Hiram Maxim's brother and son were both inventors. His brother, Hudson Maxim, worked in the field of explosives, and from 1885 the brothers worked together on the invention of smokeless gunpowder, but it was Hiram who, in 1888, perfected it and who is named on the patent. In 1905 his son, Hiram Percy Maxim, invented the Maxim silencer (for guns) as a development of his work on car silencers—Percy described his father as a cruel eccentric, but Maxim preferred to call himself a "chronic inventor."

Maxim once said: "If a domestic goose can fly, so can a man," and between 1889 and 1894 he developed a steam-powered aircraft, which succeeded in lifting off the ground, but crashed almost immediately afterward on its trial in Bexley, Kent, England.

BOUNCING BOMB

The bouncing bomb, also known as the Dambuster bomb, was invented by aeronautical engineer Barnes Wallis as a means of destroying a number of tactically important dams in Germany's Ruhr Valley during the Second World War.

It is not true that Barnes Wallis (England) thought of the notion to destroy German dams during the Second World War, but he did invent the bomb that enabled the RAF to do so. The British government was aware of the importance of the Ruhr Valley dams and, as early as 1938, the Air Ministry had discussed the possibility of destroying them if war broke out. The main objectives would be to cut off industrial and domestic water supplies and to cause flooding and damage to industrial plants and transport networks in the valley.

The prime targets were identified as the Möhne, Eder, and Sorpe dams, but the plan was shelved because of the perceived impossibility of attacking and destroying such structures—a dam wall is an extremely narrow target and, even if hit, an explosion at the top of the wall would be largely ineffective. Conventional bombing simply would not work, so Wallis invented a bomb that would bounce across the surface of the reservoir, hit the dam wall, and sink to the most vulnerable point at the foot of the dam before exploding. The key to making the bomb bounce was, in the words of Wallis's patent, to "impart a spinning motion thereto about a horizontal axis at right angles to the direction of attack." The patent drawings show that Wallis envisaged using this principle not only to attack dams but also ships in shallow water: The bomb would bounce across the water, hit the side of the ship, and then sink beneath the hull, where it would do more damage, before exploding.

Wallis eventually convinced the Air Ministry that his idea could work (although not against the type of construction used for the Sorpe dam), and on March 31, 1943, the specially formed 617 Squadron, under the command of Wing Cmdr. Guy Gibson (England), began training to fly and navigate at "zero feet." After practicing over Reculver Bay in Kent with Wallis's bouncing bombs, 617 Squadron set out on the night of May 16–17, lit by a full moon, and succeeded in destroying the Möhne and Eder dams.

The inventor: Barnes Wallis

1887 Born Barnes Neville Wallis on September 26 in Ripley, Derbyshire, the son of a doctor. Later wins a scholarship to Christ's Hospital School, London

1904–08 Apprenticed as a marine engineer to Thames Engineering Company

1913 Joins the Vickers Company

1923 Appointed chief designer of structures at Vickers Aviation, for whom he designs the R100 airship, two bomber aircraft (the Wellesley and the Wellington), and the bouncing bomb

1929 R100 makes its maiden flight

1930 On July 29 R100 makes its first passenger-carrying

Below Barnes Neville Wallis in his study at Vickers Armstrong, Weybridge, Surrey. On the wall is a photograph of the Möhne dam after its destruction by his bouncing bomb (1965)

Right A plane dropping a bouncing bomb in training for the Dambuster raid in Germany (1943)

flight, from Cardington, England, to Montreal, Canada. The flight takes 78 hours 51 minutes

1936 A prototype Vickers Type 271 bomber, later known as the Wellington, makes its maiden flight, on June 15. Wallis's design incorporates a revolutionary geodetic fuselage structure, enabling it to withstand far heavier bombardment than conventional structures

1940 Wallis outlines his concept for the bouncing bomb in a paper entitled "A note on a method of attacking the axis powers"

1942 Develops the bouncing bomb and files a patent for: "Improvements in explosive missiles and means for their discharge." For security reasons the patent is not granted until 1963

1945 Elected a Fellow of The Royal Society. Invents the concept of the variable-geometry or "swing-wing" aircraft, which he calls an "aerodyne" (see 1950s below)

1945–71 Works as chief of aeronautical research and development for the British Aircraft Corporation

1950s Designs the first experimental "swing-wing"

aircraft, the Swallow. Although the project is later abandoned, some of the technology is used in the design of the American F1-11 fighter

1968 Receives a knighthood

1971 Retires

1979 Dies on October 30 in Leatherhead, Surrey, aged 92

Best known as a mathematician and the inventor of logarithms, John Napier, surprisingly for an academic and strict Presbyterian, was also the inventor of a number of war machines, including a primitive precursor of the tank.

LOGARITHMS

John Napier (Scotland) was born in 1550, but it is uncertain whether in Balfron, Stirlingshire, or at the family seat of Merchiston Castle in Edinburgh, which today is part of the campus of Edinburgh's Napier University. Although he was an accomplished mathematician, Napier left St. Salvator's College, St. Andrews, at the age of 13 in order to travel around continental Europe. He then returned to Scotland (where he became Eighth Laird of Merchiston at the age of 18) to devote himself to an academic life writing religious works and studying literature and science. For defense against Philip II of Spain, Napier invented several war machines including an armored chariot, but the defeat of the Spanish Armada in 1588 meant that these inventions were never put to the test.

From about 1590 Napier began to investigate the concept that multiplication and division could be performed as a series of additions and subtractions, and he invented a system of logarithms to facilitate this idea. It is known that Justus Byrgius (Switzerland, aka Joost or Jost Bürgi) was also working independently on the idea of logarithms, but Napier was the first to perfect a system, which he published in 1614. Napier's logarithms, which he had formulated to the base e, were simplified in 1617 by Henry Briggs (England), after consultation with Napier. Briggs proposed using base 10 rather than base e, thus creating decimal logarithms, aka Briggsian logarithms. That same year, Napier published details of a calculating aid he had invented, which became known as "Napier's Bones" and comprised a series of cylindrical rods or bones, inscribed with numbers, which could be used to calculate logarithms.

Logarithms proved to be a vitally important advance in mathematics, and not just as a theory for academics—the concept led directly to the invention of the slide rule by William Oughtred (England) c. 1622, and paved the way for mechanical and (much later) electronic calculation and computation.

The inventor: John Napier

1550 Born in Scotland, either in Balfron, Stirlingshire, or at the family seat of Merchiston Castle, Edinburgh

1580s Invents several war machines for defense against Spain, including two types of "burning mirror," intended to be used for setting fire to ships at a distance, and an armored chariot that was effectively a primitive form of tank

c. 1590 Begins work on the concept of logarithms, which he at first calls "artificial numbers"

1593 Publishes the religious work *Plaine Discouery of the whole Reuelation of Saint John*

1597 Granted a patent for the invention of a hydraulic screw controlled by a revolving axle, used as a pump to control the water level in coal mines

1608 Moves into Merchiston Castle on the death of his father

Right Napier's Rods, aka Napier's Bones, a wooden box containing inscribed rotating cylinders that could be used to calculate logarithms

Opposite top Scottish mathematician John Napier (1550-1617) came up with the concept of logarithms in the 1590s, to the aggravation of schoolchildren ever since

Did you know?

Justus Byrgius, who invented logarithms independently of, but later than, Napier, was also a horologist. Circa 1577, he invented the first clock known to have had a minute hand (earlier clocks had only an hour hand), which he built for the Danish astronomer Tycho Brahe. Byrgius later assisted German astronomer Johannes Kepler, for whom he invented the celestial globe.

1614 Publishes his system of logarithms in *Mirfici Logarithmorum Canonis Descripto,* or "Description of the Magnificent Canon of Logarithms"

1616 & 1617 Meets Henry Briggs (England) and accepts Briggs's proposal to simplify logarithms by formulating them to base 10

1617 Publishes details of Napier's Bones, his calculating aid, in *Rabdologiae,* or "A Study of Divining Rods." Dies on April 4 at Merchiston Castle, Edinburgh

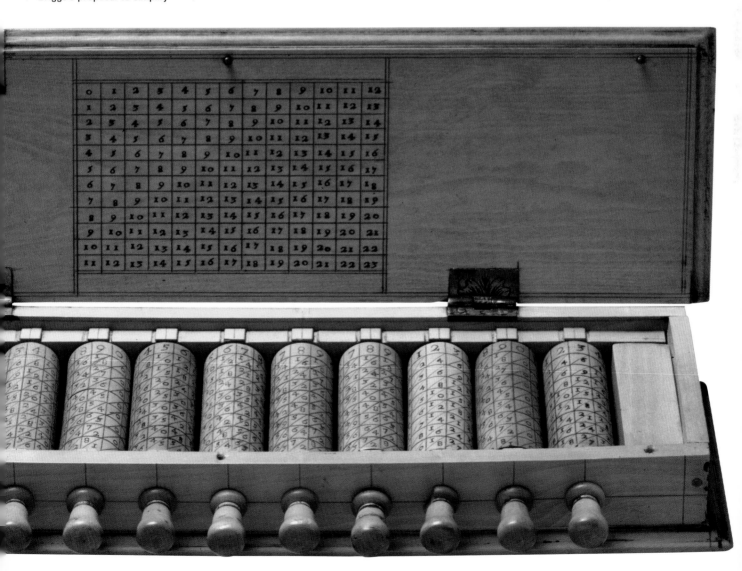

Daytime Readout Universal Imaging Device may sound like an extremely fancy name for a sundial, but the DRUID is no ordinary sundial. Invented in 1987 by John Singleton, it is the world's only sundial to cast a "smart shadow."

DRUID

Did you know?
According to family lore, Singleton, whose full name is John Stephenson Singleton, is a direct descendant of railway pioneer George Stephenson. As well as the DRUID, John has invented a number of perpetual calendars and several novel clock mechanisms that drive moon dials.

First there was the shadow clock, which indicated the passage of time using the shadow of a rod, or gnomon, placed in the ground. Later, c. 700 B.C., the gnomon came to be used in conjunction with a scale, and the sundial was born. Despite its shortcomings (a non-linear scale and a nonadjustable, self-shading gnomon with a limited hour span), there was little further progress in sundial technology until 1987, when John Singleton (England) invented the DRUID spiral sundial, which overcame all these problems.

Singleton was inspired to invent the DRUID after reading a book on sundials. He identified the failings of existing devices, listed above, and realized that they could all be overcome if he replaced the dial with a spiral scale. He showed his brother Barrie a tiny wire and paper model of the spiral sundial and Barrie was immediately eager to become John's development engineer: "I returned triumphantly with a spiral cut from plastic pipe mounted on a dowel rod with spokes. John deflated me by pointing out that the rod shadow was too wide for accurate reading of the time from its shadow. That night he conceived the idea of making the rod/shadow a precise unit of time, and the greatest advance in a millennium was achieved."

The great advance was the "smart shadow," achieved by making the rod such that it cast a shadow with a width equating to exactly half an hour, with the correct time at the center of the shadow. The numerals on the spiral also had a width of half an hour, made up of three sections equating to ten minutes each, so by relating both edges of the shadow to the numerals, the DRUID could easily be read to within five minutes' accuracy. Not only that, but it could be adjusted for local latitude by tilting the entire structure within the half-moon frame, or for seasonal hour changes by turning the spiral on its axis. A truly universal sundial, Singleton's ingenious invention was marketed in 1991 with the phrase "If you want to know the time, ask a DRUID."

Left DRUID **Above** John Singleton (right) with brother Barrie

The inventor: John Singleton

1932 Born John Stephenson Singleton on October 28 in Manchester, England. Later studies at Ewell Boys' School and Epsom Grammar School, Surrey. While in the Scouts, invents a "shoelace-tying shortcut"

1954 Graduates from Nottingham University with a degree in physics

1954–56 Does his National Service in Hong Kong; departs as "Roger Mike" (radio mechanic) and returns as "Romeo Tango" (radio technician)

1956–82 Works for Philips/ Mullard and is named on several of the company's patents

1957 Invents the block calendar, a perpetual calendar consisting of three numbered wooden cubes in a frame, and is granted a patent on March 17, 1958, for: "Improvements in perpetual calendar devices"

1982 Takes early retirement to buy into Tangible Industrial Technology Ltd., his brother Barrie's chemical products company, and acts as accountant

1987 Invents the DRUID and files a patent on November 19 (granted May 1, 1991)

1999 Invents a direct acting adjustment device so that the DRUID can be adjusted to accommodate the fact that the Earth's rotation around the sun is not uniform, which means that conventional sundials are truly accurate on only four days of the year

Did you know?

John Singleton's brother Barrie is also an inventor, having invented a filler for liquid chemicals, a wax dispersion device, a foolproof tire pressure gauge, an "incremental bender" for manufacturing the DRUID spirals, and an aqueous alternative to toilet paper.

Every living person has an appointment with death, making the subject fertile ground for inventors, and patents have been granted covering everything from the disposal of bodies to the preservation of the dead for the afterlife.

LIFE SIGNALS

For centuries most Western cultures buried their dead, but even this seemingly simple procedure presented problems. Nineteenth-century England was notorious for the activities of so-called resurrectionists, who would dig up bodies from cemeteries and sell them to medical schools. The problem was so widespread that in 1818 Edward Lillie Bridgman (England) invented a supposedly impregnable cast-iron coffin to deter grave robbers, although in practice Bridgman's coffins proved vulnerable to sledgehammers.

By the mid-19th century people were apparently more worried about being buried alive than about grave robbers, prompting the U.S. Patent Office to establish an entirely new classification of patents called "Life Signals." Frank Vester (U.S.) was granted the first patent in this category in 1868, and a further 23 examples were patented before 1900, most of them being variations on the theme of enabling a person buried alive to contact those on the surface. Most devices were triggered either by unconscious movement or by a deliberate signal from the buried person, but one particularly pointless version was patented by George Willems (U.S.) in 1908: His invention incorporated a periscope at the foot of the coffin, but if the occupant did regain consciousness, he or she would have to wait for someone to switch on the electric light provided and look into the periscope before any alarm could be raised.

In 1869, the year after the Life Signals category was established, L. Brunetti (Italy) invented the first incinerator to be specifically designed for cremation—an invention that certainly helped to reduce the number of people buried alive. Then, in 1903, Joseph Karwowski (Russia living in U.S.) went to the other extreme and invented a method of *preserving* the dead. Karwowski's patent described: "A means whereby a corpse may be hermetically sealed within a block of transparent glass...so that it will be prevented from decay and will at all times present a life-like appearance."

Life Signals time line

1818 Edward Lillie Bridgman (England) is granted a patent for a cast-iron coffin invented as security against resurrectionists

1868 Frank Vester (U.S.) is granted the first patent in the U.S. Patent Office category of Life Signals

1869 L. Brunetti (Italy) invents the first incinerator purposely designed for cremation

1887 Karl Redl (Austria) is granted a patent for a safety device that can be attached to coffins to provide an air supply and activate an electric bell "for the saving of buried living persons"

1892 Adalbert Kwiatkowski (Germany) invents and patents the deliverance coffin, which incorporates a spring-loaded signal to alert people at ground level and allow air into the coffin if there is any movement from a person buried alive

1896–97 Michael Karnicki (Poland) is granted four U.S. patents relating to the invention of a spring-loaded coffin with signals, including a flag, a bell, a speaking tube, and a lamp that burns after sunset to alert people aboveground

1903 Joseph Karwowski (Russia, living in U.S.) invents and patents his "Method of preserving the dead"

1908 George H. Willems (U.S.) invents and patents a: "Grave attachment," comprising a combined air tube, periscope, and electric light

1909 Angelo Raffaele Lerro (Italy, living in U.S.) invents and files a patent for a hermetically sealed burial casket (granted 1910), invented, according to the patent: "To overcome certain objections to which an ordinary burial casket or coffin is subject both from the viewpoint of sentiment [which he explains as "unpleasant associations attending the thought of decay"] and that of sanitation...[by preventing] the pollution of the soil and watercourses by the products of putrefaction"

1954 Asger Holm Bangsgaard (Denmark) invents and files a patent for a corpse detector (granted 1956), which works by detecting the gases created by putrefaction

1997 Koji Ishidate (Japan) invents and patents a sundial for a gravestone, the purpose of which, according to the patent, is: "To remind people of the history of the deceased with familiarity when they visit a grave in the equinoctial week or at the deathday by symbolizing a gravestone showing a time elapse in a sun-dial to image a time elapse in life"

1999 Emile Vilaplana (France) invents and (despite the 1997 precedent) patents a sundial for a gravestone in which a crucifix acts as a gnomon to indicate anniversaries and/or religious festivals, etc.

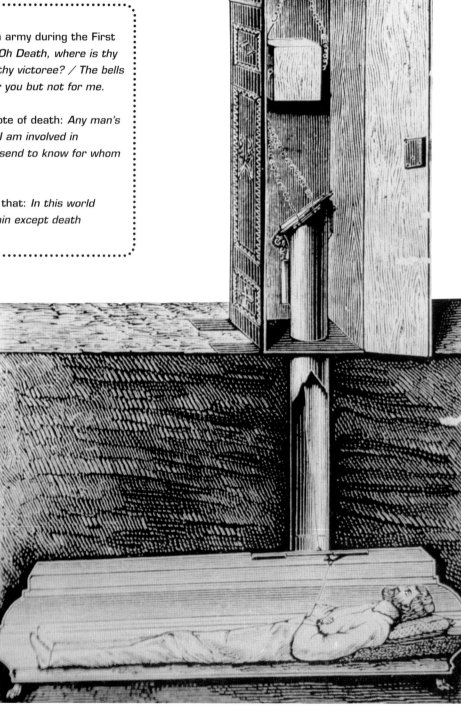

Opposite top Resurrectionists, or body snatchers, removing a body from the grave to sell for profit (1868) **Above** The patent drawing for Angelo Raffaele Lerro's hermetically sealed burial casket (1909) **Right** A German variation on the Life Signals idea. An electric wire is attached to the body by which means the slightest movement triggers an alarm bell and opens an airshaft built into the tombstone (1878)

AND ALSO...

Wind tunnel

In 1912 engineer Gustave Eiffel (France) invented an aerodynamic blast engine, a device better known as a wind tunnel. It was not the world's first wind tunnel, but it was the first to include a diffuser and a collector, setting the pattern for modern wind tunnels. Eiffel, of course, is better known for his Eiffel Tower, completed in Paris in 1899.

Rare earth magnet valve

In 1999 Ron Northedge (England), chairman of the Euromatic Machine & Oil Co., invented and filed a patent for a new type of valve for controlling the flow of liquids (patent granted 2001). The valve was opened and closed by reversing the polarity of a rare earth magnet, and was held in position magnetically without the need for springs or a continuous supply of electrical power. This enabled it to be used in a number of applications where hygiene or heat control were priorities, such as the manufacture of pharmaceuticals, drinks, and petroleum products. Reversing the polarity of the magnet required a solenoid to be energized only for a matter of milliseconds, thus the valve used a very small amount of power and could be controlled remotely.

Fuel cell

Fuel cells sound like modern alchemy. Hydrogen and oxygen are combined to create a source of power, and the only by-product is water—a concept that could potentially provide a totally clean power source for vehicles or static motors. Currently the only drawback is that most methods of producing hydrogen for use in the cell create greenhouse gases, although this is a relatively easy hurdle to overcome. The fuel cell was invented by Francis Thomas Bacon (England) of the British Electrical and Allied Industries Research Association in 1949 (patent granted 1952), but it was not until the end of the 20th century that technology had advanced far enough to manufacture a cost-effective version for general use. This was invented by Jingzhu Wei, Charles Stone, and Alfred Steck (all Canada) of Ballard Power Systems, who filed a patent in 1993 (granted 1995) for: "Trifluorostyrene and substituted trifluorostyrene copolymeric compositions and ion-exchange membranes formed therefrom." The basic principle is that a chemical reaction ionizes the hydrogen and the oxygen, which then combine to form water (H_2O) and energy—the technology comes in the catalysts and ion-exchange membranes mentioned in the patent. Early fuel cells were used in NASA's *Gemini* space missions during the 1960s, and modern fuel cells employing Ballard Power Systems technology have already successfully powered buses in the cities of Vancouver, Canada, and Chicago, U.S.

Dynamite

Dynamite was invented by Alfred Nobel (Sweden) for use in quarrying and engineering. Existing explosives such as guncotton and saltpeter were extremely volatile, so Nobel began to investigate nitroglycerine, which had been discovered by Ascanio Sobrero (Italy) in 1846. Liquid nitroglycerine was as volatile as earlier explosives, but in 1866 Nobel discovered that if it was mixed with an absorbent sand called *kieselguhr*, it could be handled safely and detonated with a blasting cap. In 1867 he filed a patent for: "Improvements in explosive compounds and in means of igniting same" and called his invention Nobel's Safety Powder, before coming up with the trademark Dynamite from the Greek *dynamis,* meaning "power." Nobel, who went on to invent gelignite in 1875, hoped that the explosive power of his inventions would be a deterrent to war, but it was not to be.

In 1888 Nobel read his own obituary in a newspaper, which had confused him with his recently deceased brother Ludwig. The obituary referred to him as "the dynamite king" and as "a merchant of death." Stung by

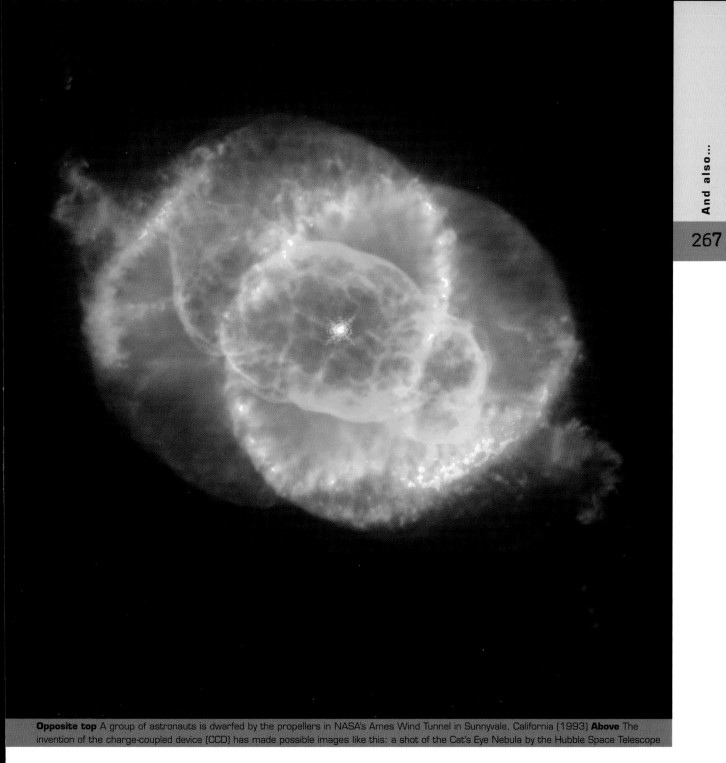

Opposite top A group of astronauts is dwarfed by the propellers in NASA's Ames Wind Tunnel in Sunnyvale, California (1993) **Above** The invention of the charge-coupled device (CCD) has made possible images like this: a shot of the Cat's Eye Nebula by the Hubble Space Telescope

the lack of reference to his philanthropic activities, he decided to establish the foundation that has awarded the Nobel prizes annually since 1901.

Charge-coupled device

The charge-coupled device (CCD) began life as a data-storage medium, was developed by NASA for the space program, and eventually found its way into everyday life at the heart of devices such as digital cameras and camcorders. The CCD was invented in 1969 by Willard Boyle and George Smith (both U.S.), who were working at Bell Laboratories where, according to Smith, they

"started batting ideas around and invented charge-coupled devices in about an hour." A CCD is a silicon chip containing a number of light-sensitive cells called pixels, which convert light into an electrical signal. NASA saw the potential of this invention for imaging, rather than for data storage as originally intended by the inventors, and financed further development to the point where imaging CCDs were advanced enough to be used in the Hubble Space Telescope. The first imaging CCD, produced in 1974, had a mere 1,000 pixels—a far cry from the hundreds of thousands of pixels that appear in even the simplest of today's digital cameras.

Inventions without wings is a collection of inventions that did not, or have not yet, become a practical reality. These include fictional inventions, fantasy inventions, and some that are destined never to go into production.

INVENTIONS WITHOUT WINGS

Method of growing unicorns

On February 7, 1984, Timothy G. Zell (U.S.) was granted a patent for a surgical means of growing unicorns. Zell's patent states that: "This invention relates to a method of growing unicorns in a manner that enhances the overall development of the animal"—although it might not feel like enhancement if you're a baby cow, antelope, sheep or goat undergoing Zell's procedure.

Zell's patent cites a 1936 report by biologist W. Franklin Dove (U.S.) stating: "Unicorns have been developed by a surgical procedure in which the horn buds of a newly born animal are transplanted from the usual location to a central position on the front of the animal's skull. It is not generally known that during the first week of development, the horn buds are attached to the skin only and attachment to the skull begins after this period. In other words, horns are the result of separate ossifications that subsequently fuse to the frontal bones of the head rather than being outgrowths of these frontal bones or of the skull." Zell then goes on to describe an "improved method of forming a unicorned animal" by removing the horn buds of a two-horned animal and transplanting them "adjacent to one another over the pineal gland at the front of the skull. Thereafter the resulting horns grow as one and connect with the frontal portion of the skull directly over the pineal gland to render a unicorn of higher intelligence and physical attributes."

Horse-powered road vehicle

Inventors are often accused of putting the cart in front of the horse, but on April 29, 1981, P.A. Barnes (Britain) was granted a British patent for an invention that put the horse inside the cart. The vehicle envisaged by Barnes was powered by a horse walking on an endless conveyor belt in the center of the vehicle, which acted as a treadmill and drove the wheels via a chain, a clutch, and a variable-ratio gearbox. Barnes claimed:

"There are several advantages in taking the horse off the road surface in vehicle propulsion. The most obvious is speed variation....By selecting the lowest gear the vehicle moves forward slower than the walking speed of the horse; this helps it to pull a load uphill. By selecting the highest gear the vehicle moves faster than the walking speed of the horse and so shortens the journey time." Barnes even proposed a dashboard instrumentation displaying "horse temperature and collar-push," and the patent also stated: "Containers are provided to collect dropping."

Ladder to the stars

The idea of a ladder to the stars is not a new one—in the Bible, Jacob has a vision of a ladder or staircase leading to heaven; Led Zeppelin sang about a stairway to heaven; Bob Dylan sang about a ladder to the stars, and, in his 1979 novel *Fountains of Paradise*, Arthur C. Clarke wrote about an elevator into space. Clarke said that his idea would eventually be built, but only "about 50 years after everyone stops laughing." In fact, only 24 years after he suggested it, NASA has started researching a means of building a space elevator, and in 2003 some 70 scientists and engineers convened to discuss the idea; Clarke addressed the conference by satellite link from his home in Sri Lanka.

Advances in materials science have turned Clarke's idea from science fiction into science probability, and NASA hopes to create a vertical track reaching 60,000 miles (100,000 km) into space; the designs are already under way, but the technology to build it is still a long way off. The crucial breakthrough has been the development of microscopic tubes of carbon known as carbon nanotubes, which are as strong as diamond but also flexible, which means that potentially they could be spun into a fiber. If and when scientists find a way of forming

Opposite Unicorn goat, California

such a fiber, the resulting carbon nanotube "ribbon" will in theory be strong enough to build such a track. One end would be tethered to a base station on Earth and the other would float deep in space, the vital requirement being that the center of gravity of the track would be in a geostationary orbit in order to keep the whole thing vertical. The track would be deployed by placing a satellite in a geostationary orbit; one end of the nanotube ribbon would be lowered to Earth from the satellite, the other released into space, and, once in place, the track would literally act as a ladder for satellites and spacecraft to climb into space.

Matter transporter

Gene Roddenberry (U.S.) and his co-creators of *Star Trek* invented a concept that has exercised the imagination of Trekkies and scientists alike for four decades since the first utterance of the immortal phrase: "Beam us up, Scotty." A *Star Trek* technical manual explains that the matter transporter (which transports people or objects almost instantaneously over vast distances) comprises four essential elements: a Scanner, which creates a precise record of the arrangement of all the molecules and atoms in the item or person to be transported; an Energizer, which converts the molecules into a stream of energy (hence Captain Kirk's familiar command, "Energize!"); a Pattern Buffer, which maintains the pattern of the energy stream, and an array of Emitter Pads to actually transmit the energy stream to its destination.

Unlike Arthur C. Clarke's space elevator, the chances of matter transportation (aka teleportation) becoming a reality range from "extremely low" to "zero," according to most scientists. However, researchers at three separate institutions—California Institute of Technology (Caltech), Århus University, Denmark, and the University of Wales— have independently succeeded in exploiting the phenomenon of "quantum entanglement" to remotely alter the nature of one entangled photon by physically altering the nature of its twin. (Quantum entanglement means that two entangled photons will always be mirror images of each other, no matter how far apart they are—if one alters, so does the other, something that Albert Einstein called "spooky action at a distance.")

In itself, this does not constitute teleportation because it involves the alteration of a remote object rather than the transportation of a given object to another place. However, Caltech's scientists took the process a step further: By attaching a nonentangled photon to an entangled one, they managed to create a replica of the nonentangled photon attached to the entangled twin. This has huge implications for the future

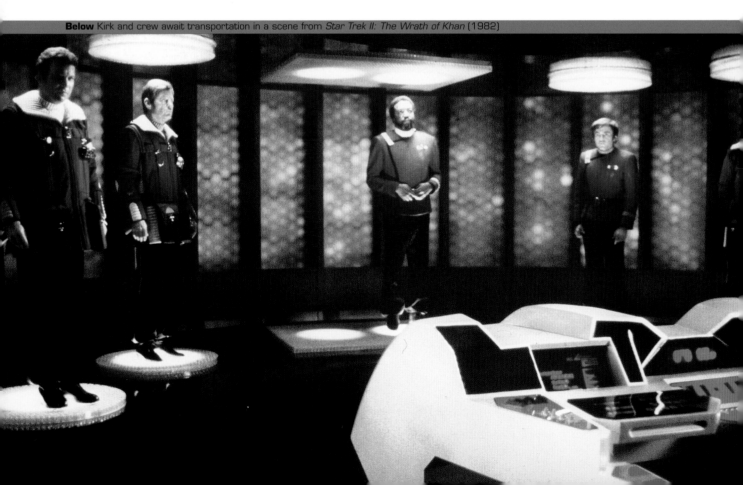

Below Kirk and crew await transportation in a scene from *Star Trek II: The Wrath of Khan* (1982)

Above Rod Taylor as the Time Traveler in a scene from *The Time Machine* (1960)

of communications, but it is still not the same as transporting Kirk from one place to another: The creation of a replica in a remote place, clever though it may be, is more like faxing a copy of something to that place than transporting the original—at present it can only be done with photons, but it is possible to fax them to remote parts of the universe.

Time Machine

In 1895 H.G. Wells (England) published his novel *The Time Machine*. The novel is an allegory describing the two-tier society that the protagonist (known only as the Time Traveler) encounters when he travels to the year 802701. The naïve Eloi live a life of seeming ease above ground, but are exploited by the brutish Morlocks, who have developed a separate society below ground, providing a narrative opportunity for Wells to explore his political interests, which included progressive education, world government (he was active in promoting the idea of a League of Nations), and human rights. Wells was also interested in scientific theory, and in order for his hero to act as a contemporary observer of a future society, Wells had to invent a time machine for him to travel in,

although the science behind it is not explained—the Time Traveler merely pushes a lever, and time speeds up until night and day are a blur.

Much has been written about time travel since, both in science journals and in fiction, but it should be remembered that *The Time Machine* was published ten years before Einstein proposed the theory of special relativity in 1905 (the more familiar theory of general relativity, $E=mc^2$, was published in 1916). These theories propose that time is not constant but depends on the speed of the traveler, and it has since been proved that time does in fact pass more slowly the faster an object travels. According to the theory, time would stop for a body traveling at the speed of light (if that were physically possible); Einstein's interpretation was that this meant it was impossible to travel faster than the speed of light, but others have taken it to imply that if it were possible to travel at such a speed, time would move backward.

Other theories about time travel include the possibility of traveling through "wormholes" in the fabric of space-time, but all forms of time travel run up against the philosophical paradox that traveling into the past could alter the future. This paradox is memorably described in

Ray Bradbury's story "A Sound of Thunder," in which a time traveler steps on an ancient butterfly and alters the future course of history, so that he returns to a present subtly different from the one he left.

TARDIS

TARDIS is an acronym for Time And Relative Dimension In Space, and is a highly sophisticated invention created by the BBC as a combination of time machine and matter transporter. Dr. Who's TARDIS may have looked like an old police telephone box, but in fact, as the name suggests, it enabled him to travel to any time, any place, anywhere, while acknowledging the existence, if not the limitations, of Einstein's theory of relativity.

For those not in the know, Dr. Who is a Gallifreyan Time Lord; Time Lords use various types of TARDIS to travel the universe (Dr. Who's is a Type 40) and, in order to remain inconspicuous, these machines have a "chameleon circuit" that enables them to appear in whatever form best suits the surroundings in which they materialize. However, the chameleon circuit in Dr. Who's TARDIS jammed after it materialized on Earth in 1963, in the days before policemen carried radios and when blue police telephone boxes were a common sight. Apart from being able to travel through time and space (the science behind this is not explained), the clever thing about the TARDIS is that it is bigger on the inside than on the outside. This is because the inside occupies a different dimension relative to the outside—just as a large box placed far away will appear small enough to fit inside a smaller box nearby.

The idea of the TARDIS was originally suggested by BBC producer Verity Lambert (England), and her concept was realized by Anthony Coburn (Australia), who came up with the idea of it being a police box, and Peter Brachacki (born Poland), who designed the console room—although Gallifreyans still labor under the illusion that time travel and the TARDIS were invented by the Time Lords Rassilon and Omega.

Heath Robinson

William Heath Robinson (England) was not an inventor, but in the course of satirizing the machine age he devised many weird and wonderful contraptions, and gave his name to any absurdly complex and dilapidated contraption designed to perform a very simple operation. Born in 1872, Robinson trained at the Islington School of Art and the Royal Academy, both in London, England, and became an artist, cartoonist, and illustrator. He is most famous for his fantastic designs for inventions that took the execution of everyday tasks, such as laying the table or shuffling cards, to ludicrous extremes; these designs gave rise to the term "Heath Robinson" or "Heath Robinson contraption" to describe any convoluted invention that makes a job more complicated than it was to begin with.

Robinson's targets for ridicule included the First World War, golfers, life in high-rise blocks, gardening, driving, and other forms of transportation, and married life. Several collected editions of his inventive cartoons were published, including *Hunlikely!* (1916), *The Homemade Car* (1921), *Humours of Golf* (1923), *Absurdities* (1934), *Railway Ribaldry* (1935), *How to Live in a Flat* (1936), *How to be a Perfect Husband* (1937), *How to Make a Garden Grow* (1938), and *How to be a Motorist* (1939). Many of his drawings were reprinted in the 1970s and 1980s in compilations, one of which was simply entitled *Inventions*.

See also: Golfing inventions, page 169; Geiger counter, page 242

Arthur Pedrick

Arthur Paul Pedrick (England) has been referred to as the king of wacky patents. After a career as a patent examiner for the U.K. Patent Office, Pedrick clearly thought that he could do better than some of the inventions that had passed before his professional eye, and spent the 1960s and 1970s filing a vast number of patents, none of which was a commercial success and many of which defied the laws of physics.

Among his patented inventions are: a golf ball with aerodynamic flaps to control spin; a golf tee incorporating photoelectric cells to prevent mishits; a means of solving world famine by piping snowballs from Antarctica to irrigate the deserts of Australia, and another means of solving world famine by piping fresh water from the Amazon to irrigate the Sahara, using a pipeline stretching across the Atlantic from South America to Africa. As well as world famine, Pedrick was also concerned with world peace, and in 1974, during the Cold War, he patented the idea of placing three satellites in geostationary orbits over Washington, Moscow and Peking as a deterrent to war—the satellites would be programmed to drop nuclear bombs on all three cities if one of the superpowers attacked either of the others.

See also: Golfing inventions, page 169; Geiger counter, page 242; Magic eye, page 234

Right "For Green Winters in Switzerland—A Simple Device for Making your Own Snow as You Go Along," undated W. Heath Robinson drawing

PATENT NUMBERS LIST

The year given is the date of invention (where known), and is not necessarily the same year a patent was filed or granted; where the year of invention is not known, the year given is the date the patent was filed; date formatting is as recorded in the Patent Office. "Nationality" refers to the nationality of the inventor, and is not necessarily the country in which the invention was conceived or patented.

AC induction motor 1887
Nikola Tesla (Croatia–USA)
Patent no: US 381,968
(filed 1887, granted 1888)

Abrasive shaving device 1899
Samuel L. Bligh (USA)
Patent no: US 729,083
(filed 1899, granted 1900)

Actar 911 CPR mannequin 1989
Dianne Croteau et al (Canada)
Patent no: US 4,984,987
(filed 1989, granted 1991) &
WO 91/07737

Action Man Dynamic Physique 1977
William A. G. Pugh et al (Britain)
Patent no: GB 48,465/77 &
US 4,274,224 (filed 1978,
granted 1981)

Aerodyne (swing-wing aircraft) 1945
Barnes Neville Wallis (England)
Patent no: US 2,915,261
(filed 1954, granted 1959)

Aerosol can (first) 1862
J.D. Lynde (USA)
Patent no: US 34,894

Aerosol can (modern) 1926
Erik Rotheim (Norway)
Patent no: Norwegian patent
46,613 & US 1,800,156 et al

Air brakes (railway) 1869
George Westinghouse, Jr (USA)
Patent no: US 88,929 (reissued
1874 as Re 5,504) &
GB 1,691/1872

Air conditioning 1902
Willis H. Carrier (USA)
Patent no: US 808,897
(filed 1904, granted 1906)

Anglepoise lamp (first) 1932
George Cawardine (England)
Patent no: GB 404,615

Anglepoise lamp (3-spring) 1934
George Cawardine (England)
Patent no: GB 433,617

Anywayup cup 1990
Mandy Haberman (England)
Patent no: GB-B-2266045

(filed & granted 1992)

Artificial heart (first to be permanently implanted) 1977
Robert K. Jarvik (USA)
Patent no: US 4,173,796
(filed 1977, granted 1979)

Artificial respirator (practical) 1927
Philip Drinker & Louis Agassiz Shaw (USA)
Patent no: US 1,834,580
(filed 1929, granted 1931)

Automated tea-maker (first practical example) 1902
Frank Clarke (England)
Patent no: GB 15,170/1902

Baby patting machine 1968
Thomas V. Zelenka (USA)
Patent no: US 3,552,388
(filed 1968, granted 1971)

Ballpoint pen (concept) 1888
John J. Loud (USA)
Patent no: US 392,046 &
GB 15,630/1888

Ballpoint pen (practical) 1943
Laszlo Josef Biró
(Hungary–Argentina)
Patent no: GB 564,172
(filed 1943, granted 1944) &
US 2,390,636

Bar code 1949
Norman Woodland & Bernard Silver (USA)
Patent no: US 2,612,994
(filed 1949, granted 1952)

Barbed wire (first patent) 1867
Lucien B. Smith (USA)
Patent no: US 66,182

Barbed wire (commercially successful) 1873
Joseph F. Glidden (USA)
Patent no: US 157,124
(filed 1873, granted 1874)

Beta-blockers 1962
James Black (Scotland)
Patent no: GB 909,357 (for
pronethalol) & GB 994,918
(for propranolol)

Bicycle hub gears 1901
John James Henry Sturmey
(England)
Patent no: GB 16,221/1901

Biogun 1994
Jonathan Copus (England)
Patent no: GB 2,246,955

Bionic ear 1978
Graeme Clark et al (Australia)
Patent no: AU 529,974
(filed 1978, granted 1980)

Bouncing bomb 1940
Barnes Neville Wallis (England)
Patent no: GB 937,959
(filed 1942, granted 1963)

Bowden cable 1896
Ernest Mannington Bowden
(Britain)
Patent no: GB 25,325/1896

Breakfast cereal (Shredded Wheat, the first breakfast cereal) 1893
Henry D. Perky (USA)
Patent no: US 502,378

Breakfast cereal (Granose Flakes, the first flaked breakfast cereal) 1894
John Harvey Kellogg (USA)
Patent no: US 558,393
(filed 1895, granted 1896)

Breakfast cereal (puffed cereals) 1902
Alexander Pierce Anderson (USA)
Patent no: GB 13,353

Camera (roll-film) 1888
George Eastman (USA)
Patent no: US 388,850 &
GB 6,950/1888

Camera track and controls (for televising sports events) 1998
Matthew Allwork (England)
Patent no: WO 98/306,694 &
AUS 881,838/98

Can opener (first) 1855
Robert Yeates (England)
Patent no: GB 1577/1855

Can opener (first with cutting wheel) 1870
William W. Lyman (USA)
Patent no: US 105,583

Cans for food 1810
Peter Durand (England)
Patent no: GB 3,372/1810

Carbon paper 1806
Ralph Wedgwood (England)
Patent no: GB 2972/1806

Cash register 1879
James Ritty & John Ritty (USA)
Patent no: US 221,360

CAT scanner (computer-aided tomography) 1968
Godfrey Hounsfield (England)
Patent no: GB 1,283,915 &
US 3,778,614

Cathode ray direction finder 1926
Robert Watson-Watt (Scotland)
Patent no: GB 252,263

Catseyes 1935
Percy Shaw (England)

Patent no: GB 457,536
(filed 1935, granted 1936)

Cavity magnetron (re: microwave oven) c.1940
John Randall & H.A.H. Boot
(England)
Patent no: GB 588,185 (1947)

Celluloid roll film 1887
Hannibal Goodwin (USA)
Patent no: US 610,861
(filed 1887, granted 1898)

Chainsaw (petrol driven) 1927
Emil Lerp (Germany)
Patent no: GB 328,527
(granted 1930)

Charge-coupled device 1969
Willard Boyle & George Smith
(USA)
Patent no: US 3,796,927
(filed 1970, granted 1974) &
GB 1,358,890 (filed 1971,
granted 1974)

Clasp locker (early form of zip) 1891
Whitcomb L. Judson (USA)
Patent no: US 504,037-8
(filed 1891, granted 1893)

Cold air blast wake-up apparatus 1976
Peter MacNeil (Canada)
Patent no: US 4,031,711 (filed
1976, granted 1977)

Computer (stored programme) 1948
Professor Freddie Williams &
Tom Kilburn (England)
Patent no: GB 657,591

Computer mouse 1967
Douglas Engelbart (USA)
Patent no: US 3,541,541
(filed 1967, granted 1970)

Concealed liquor flask 1885
Herbert William Torr Jenner
(Britain)
Patent no: US 330,709 &
GB 14,055/1885

Corpse detector 1954
Asger Holm Bangsgaard
(Denmark)
Patent no: GB 753,865
(filed 1954, granted 1956)

Cotton buds 1925
Leo Gerstenzang (Poland–USA)
Patent no: US 1,721,815
(filed 1927, granted 1929)

Crown cork 1891
William Painter (USA)
Patent no: US 468,226 &
GB 2031/1892 (both filed 1891,
granted 1892)

Dehydrated mashed potato 1961
Edward A. M. Asselbergs et al (Canada)
Patent no: CA 816,400 &
US 3,260,607 (granted 1966) &
GB 965,586

Dishwasher 1885
Josephine Cochran (USA)
Patent no: US 355,139
(filed 1885, granted 1886) &
GB 9895/1887

Disposable diaper c.1949
Marion Donovan (USA)
Patent no: US 2,575,163-5

DRUID spiral sundial 1987
John Singleton (England)
Patent no: GB 2,212,630
(filed 1987, granted 1991),
Canada 1,313,756 (granted
1993) & US 4,922,619

**Dual Cyclone vacuum cleaner
1978**
James Dyson (England)
Patent no: US 4,373,228
(filed 1979, granted 1983)

Dynamite 1866
Alfred Nobel (Sweden)
Patent no: GB 1,345/1867 &
US 78,317 (US granted 1868)

Elastic band 1845
Stephen Perry (England)
Patent no: GB 10,568/1845

**Electric food mixer (blender)
1922**
Stephen J. Poplawski (USA)
Patent no: US 1,480,914
(filed 1922, granted 1924)

**Electric food mixer (multi-
purpose) 1950**
Ken Wood (England)
Patent no: GB 664,634

Electric iron 1882
Henry W. Seely (USA)
Patent no: US 259,054

Epsom salts 1698
Nehemiah Grew (Britain)
Patent no: GB 354 (1698)

**Facsimile machine (concept)
1843**
Alexander Bain (Scotland)
Patent no: GB 9,745/1843

**Facsimile machine (practical)
1861**
Giovanni Caselli (Italy)
Patent no: GB 2,395/1861

Film roll holder 1885
George Eastman & William H.
Walker (USA)
Patent no: US 316,952

Fuel cell (first) 1949
Francis Thomas Bacon (England)
Patent no: GB 667,298
(filed 1949, granted 1952)

**Fuel cell (commercially
practicable) 1993**
Jingzhu Wei et al (Canada)

Patent no: WO 95/08581
(filed 1993, granted 1995) &
US 5,422,411

G.I. Joe (Action Man) 1964
Hassenfeld Brothers Inc. (USA)
Patent no: US 3,277,602
(granted 1966)

Gelignite 1875
Alfred Nobel (Sweden)
Patent no: GB 4,179/1875

**Geostationary orbiting bombs as
nuclear deterrents 1974**
Arthur Paul Pedrick (England)
Patent no: GB 1,361,962

Golf ball (rubber core) 1898
Coburn Haskell & Bertram George
Work (USA)
Patent no: GB 17,554/1898

**Golf ball (machine for winding
rubber core) 1900**
W. Millinson for Haskell Golf Ball
Co. (USA)
Patent no: GB 4,165/1900

Golf ball (walking) 1968
Donald B. Poynter (USA)
Patent no: US 3,572,696
(filed 1968, granted 1971)

**Golf balls, improvements in the
flight direction and location of
1967**
Arthur Paul Pedrick (England)
Patent no: GB 1,121,630
(filed 1967, granted 1968)

Golf club (steel-shafted) 1910
Arthur F. Knight (USA)
Patent no: US 976,267

Golf club (breakable) 1960
Ashley Pond III (USA)
Patent no: US 3,087,728
(filed 1960, granted 1963)

Golf swing trainer (aquatic) 2001
John Carr (USA)
Patent no: US 6,325,727

Golf tee 1899
George F. Grant (USA)
Patent no: US 638,920

**Golf tee (with photoelectric cells
and compressed air to prevent
mishits) 1970**
Arthur Paul Pedrick (England)
Patent no: GB 1,251,780
(filed 1970, granted 1971)

Gramophone 1887
Emile Berliner (Germany–USA)
Patent no: US 372,786

Haberman Feeder 1982–84
Mandy Haberman (England)
Patent no: GB 2169210 &
2131301

Helmet camera 1998
Matthew Allwork (England)
Patent no: WO 98/301,390

**Holder for electroplating
(Edison's last patent) 1931**
Thomas Alva Edison (USA)

(patent granted 1933)

Hook-and-eye fastener 1843
Charles Atwood (USA)
Patent no: US 2,978

Hose clip (Jubilee Clip) 1920
Lumley Robinson (England)
First patent filed 1920, granted
1921, superseded by
GB 293,279 (1928)

Hovercraft 1955
Christopher Cockerell (England)
Patent no: GB 854,211
(filed 1955, granted 1960)

Incubator 1880
Odile Martin (France)
Patent no: FR 136,015 &
US 237,589 & GB 4308/1880

**Inseparable tear-strip drinks can
(precursor of 'push-in fold-back')
1973**
Omar Brown for Ermal Fraze
(USA)
Patent no: US 3,870,001
(filed 1973, granted 1975)

Insulin 1921
Frederick Banting et al (Canada)
Patent no: CA 234,336 &
GB 203,778 & US 1,469,994
(various dates)

Jet engine 1929
Frank Whittle (England)
Patent no: GB 347,206
(filed 1930, granted 1931) &
US 2,404,334 (filed 1941,
granted 1946)

**Jet Ski (personal watercraft)
1965**
Clayton Jacobson II (USA)
Patent no: US 3,369,518
(filed 1966, granted 1968)

Kevlar 1963
Stephanie Kwolek (USA)
Patent no: US 3,287,323
(filed 1963, granted 1966)

**Kinetograph & Kinetoscope
(motion picture camera and
peephole viewer) 1891**
Thomas Alva Edison (USA)
Patent no: US 493,426
(granted 1893)

Land Shark 1997
David Baker (England)
Patent no: US 6,386,929
(filed 1997, granted 2002)

**Landlord's Game (precursor of
Monopoly) 1903**
Lizzie J. Magie (USA)
Patent no: US 748,626
(filed 1903, granted 1904)

**Landlord's Game (improved
version) 1923**
Lizzie J. Magie Phillips (USA)
Patent no: US 1,509,312
(filed 1923, granted 1924)

Laser (theoretical) 1951
V.A. Fabrikant et al (USSR)
Patent no: USSR 123,209

(filed 1951, granted 1959)

**Laser (claim for precedence)
1957**
Gordon Gould (USA)
Patent no: US 4,053,845
(filed 1959, granted 1977)

Laser (first practical) 1960
Theodore H. Maiman (USA)
Patent no: based on Schawlow &
Townes' US 2,929,922 (1958)

Lava lamp 1963
Craven Walker (England)
Patent no: GB 1,034,255
(filed 1964)

Light bulb 1879
Joseph Swan (England)
Patent no: GB 4,933/1880

Light bulb 1879
Thomas Edison (USA)
Patent no: US 223,898
(filed 1879, granted 1880,
rescinded 1883)

Logarithms 1614
John Napier (Scotland)
Published 1614 in *Mirfici
Logarithmorum Canonis
Descripto*, or '*Description of the
Magnificent Canon of Logarithms*'

Logmaster clamp 1987
Charles Brathwaite (Barbados)
Patent no: WO 02/34488
(filed 2000, patent pending)

Lorenzo's Oil 1989
Michaela & Augusto Odone (USA)
Patent no: US 5,331,009 (filed
1989, granted 1994)

Machine gun (first) 1718
James Puckle (England)
Patent no: GB 418

**Machine gun (first successful)
1861**
Richard Gordon Gatling (USA)
Patent no: US 36,836 (granted
1862) & GB 790/1865

Machine gun (automatic) 1883
Hiram Stevens Maxim
(USA–Britain)
Patent no: GB 3,493/1883 &
US 317,161

**Magic cube 1972 (see Rubik's
Cube**
Larry Nichols (USA)
Patent no: US 3,655,201

Manner of buoying vessels 1849
Abraham Lincoln (USA)
Patent no: US 6,469

Meccano 1900
Frank Hornby (England)
Patent no: GB 587/1901 (filed &
granted 1901)

**Method of cutting shirt patterns
1848**
Oliver Winchester (USA)
Patent no: US 5,421

Microchip (first integrated circuit) 1958
Jack Kilby (USA)
Patent no: US 3,138,743 et al &
GB 945,734 et al (filed 1959,
granted 1964)

Microchip (planar integrated circuit, first commercially available integrated circuit) 1959
Robert Noyce et al (USA)
Patent no: US 3,117,260
(filed 1959, granted 1964)

Microwave oven 1945
Percy LeBaron Spencer (USA)
Patent no: US 2,495,429
(filed 1945, granted 1950)

Miniature electronic calculator 1967
Jack Kilby et al (USA)
Patent no: US 3,819,921
(granted 1974)

Monopoly 1933
Charles B. Darrow (USA)
Patent no: US 2,026,082
(filed & granted 1935)

Monopoly (London version) 1935
John Waddington Ltd. (England)
Patent no: GB 453,689

Mousetrap 1894
William Hooker (USA)
Patent no: US 528,671

Mousetrap ("Little Nipper") 1899
James Henry Atkinson (England)
Patent no: GB 13,277/1899

Multi-phase electrical transmission system 1887
Nikola Tesla (Croatia–USA)
Patent no: US 382,280
(filed 1887, granted 1888)

Napier's Bones (calculating device) 1617
John Napier (Scotland)
Published 1617 in *Rabdologiae*, or
'*A Study of Divining Rods*'

Non-stick pans 1954
Marc & Colette Grégoire (France)
Patent no: GB 820,442 (filed 1955, granted 1959) French patent filed 1954 by Colette Grégoire & Georgette Wamant

Paper clip 1899
Johann Vaaler (Norway)
(German patent filed 1899, granted 1900). US 675,761
(granted 1901)

Paper clip (Gothic) 1934
Henry Lankenau (USA)
Patent no: US 1,985,866

Paper clip-making machine 1899
W.D. Middlebrook (USA)
Patent no: US 636,272

Parking meter 1932
Carlton C. Magee (USA)
Patent no: US 2,039,544

Pat on the Back Apparatus 1985
Ralph R. Piro (USA)

Patent no: US 4,608,967
(filed 1985, granted 1986)

Perpetual calendar 1957
John Singleton (England)
Patent no: GB 831,572
(granted 1958)

Perspex 1932
Rowland Hill (Britain)
Patent no: GB 395,687
(granted 1933)

Phonograph 1877
Thomas Alva Edison (USA)
Patent no: US 200,251
(filed 1877, granted 1878) &
GB 2,909/1877 (granted 1877)

Photocopier (photographic) 1903
George C. Beidler (USA)
Patent no: US 753,351
(patented as 'Camera')

Photocopier (xerography) 1938
Chester F. Carlson (USA)
Patent no: US 2,221,776 et al

Piping snowballs from Antarctica to Australia 1966
Arthur Paul Pedrick (England)
Patent no: GB 1,047,735

Piping water from the Amazon to the Sahara 1969
Arthur Paul Pedrick (England)
Patent no: GB 1,204,648
(filed 1969, granted 1970)

Pneumatic tire 1845
Robert Thomson (Scotland)
Patent no: GB 10,990 (1845)

Pneumatic tire (reinvention) 1888
John Boyd Dunlop (Scotland)
Patent no: GB 10,607/1888

Pneumatic tyre (wired) 1890
Charles Kingston Welch (England)
Patent no: GB 14,563/1890

Polariser 1932
Edwin Herbert Land (USA)
Patent no: GB 412,179 (granted 1934)

Polaroid camera 1947
Edwin Herbert Land (USA)
Patent no: US 2,543,180 &
GB 658,740 (both filed 1948, granted 1951)

Polaroid film 1947
Edwin Herbert Land (USA)
Patent no: US 2,543,181
(filed 1948, granted 1951)

Polythene 1933
Eric William Fawcett, Reginald Oswald Gibson et al (Britain)
Patent no: GB 471,590
(full specification filed & granted 1937)

Post-it Note (adhesive) 1969
Spencer Silver (USA)
Patent no: US 3,691,140 (filed & granted 1970)

Post-it Note (notepaper concept) 1977
Arthur Fry (USA)
Patent no: protected by Silver's
US 3,691,140

Pruteen 1967–71
ICI (England)
Patent no: GB 1,353,008

Quorn 1986
ICI & Rank Hovis McDougall (England)
Patent no: GB 1,346,061

Radar 1935
Robert Watson-Watt (Scotland)
Patent no: GB 593,017
(filed 1935, granted 1947)

Radio, clockwork 1991
Trevor Graham Baylis (England)
Patent no: GB 2,262,324

Radio communication (patent for practical equipment) 1896
Guglielmo Marconi (Italy)
Patent no: GB 12,039/1896 &
US 586,193 (granted 1897)

Radio communication (patent for improvements) 1900
Guglielmo Marconi (Italy)
Patent no: GB 7,777/1900 &
US 763,772

Radio communication, triode valve 1906
Lee de Forest (USA)
Patent no: US 841,387 (granted 1907) & GB 1,427/1908
(granted 1908)

Radio communication, regenerative circuit 1913
Edwin Howard Armstrong (USA)
Patent no: US 1,113,149
(granted 1914)

Radio communication, super-heterodyne circuit 1917
Edwin Howard Armstrong (USA)
Patent no: US 1,342,885
(filed 1919, granted 1920) &
GB 137,271 (granted 1920)

Radio communication, super-regenerative circuit 1921
Edwin Howard Armstrong (USA)
Patent no: US 1,424,065 &
GB 182,135 (granted 1923)

Radio communication, FM radio 1930
Edwin Howard Armstrong (USA)
Patent no: US 1,941,066-69
(filed 1930, granted 1933)

Railway carriage re-railer 1867
George Westinghouse, Jr (USA)
Patent no: US 61,967

Railway dining carriage 1869
George Pullman & Ben Field (USA)
Patent no: US 89,537

Railway hotel carriage 1869
George Pullman & Ben Field (USA)
Patent no: US 89,538

Railway signal (successive locking) 1860
Austin Chambers (England)
Patent no: GB 31/1860

Railway sleeping carriage 1865
George Pullman & Ben Field (USA)
Patent no: US 49,992

Rare earth magnet valve 1999
Ron Northedge (England)
Patent no: GB 2,356,917
(filed 1999, granted 2001) &
WO 00/20,785

Raylo (table-top railway game) 1915
Frank Hornby (England)
Patent no: GB 27,533

Refrigeration 1834
Jacob Perkins (USA)
Patent no: GB 6,662

Refrigerator (dual-hinged door) 1997
Ian Harrison (England)
Application no: GB 9722639.3
(abandoned)

Refrigerator ship 1868
Charles Tellier (France)
Patent no: FR 81,858

Repeating rifle (first successful) 1860
Benjamin Tyler Henry for
Oliver Winchester (USA)
Patent no: US 30,446

Revolver (first successful) 1835
Samuel Colt (USA)
Patent no: US Re 124
(filed 1835, granted 1836, reissued 1848) & GB 6,909
(granted 1835)

Ring-pull for tear-strip cans 1965
Omar Brown & Don Peters for
Ermal Fraze (USA)
Patent no: US 3,349,949
(filed 1965, granted 1967)

Roller skates (in-line) 1823
Robert John Tyers (England)
Patent no: GB 4,782/1823

Roller skates (modern four-wheel) 1862
James Leonard Plimpton (USA)
Patent no: US 37,305 (granted 1863) & GB 2190/1865

Rotary steam engine (Westinghouse's first patent) 1865
George Westinghouse, Jr (USA)
Patent no: US 50,759

Rotary swing (fairground wheel) 1872
Isaac Newton Forrester (USA)
Patent no: US 169,797 (filed & granted 1875)

Rubik's Cube 1974
Erno Rubik (Hungary)
Patent no: HU 170,062
(filed 1975)

Safety pin 1849
Walter Hunt (USA)
Patent no: US 6,281

Safety razor 1901
King Camp Gillette (USA)
Patent no: US 775,134-5
(granted 1904) & GB
28,763/1902

Sawmill machinery 1865
George Westinghouse, Sr (USA)
Patent no: US 48,857

Scoring board for bridge 1900
Walter Clopton Wingfield (Wales)
Patent no: GB 13,409/1900

Scrabble 1931
Alfred Mosher Butts (USA)
Patent no: Initial application
refused. Eventually patented by
James & Helen Brunot as
US 2,752,158 (filed 1954,
granted 1956) & GB 747,598

Serraglaze 1993
Peter Milner (England)
Patent no: EP 0753121
(filed 1994, granted 2001)

Serrascope 1999
Peter Milner (England)
Patent no: EP 00966290.9
(filed 2000, patent pending)

Serraview 1989
Peter Milner (England)
Patent no: EP 0402444
(filed 1989, granted 1994)

Serravista 2002
Peter Milner (England)
Patent no: PCT/GB03/000073
(filed 2003, patent pending)

**Sewing machine (chain stitch)
1790**
Thomas Saint (England)
Patent no: GB 1,764/1790

**Sewing machine (lock stitch)
1846**
Elias Howe, Jr (USA)
Patent no: US 4,750 &
GB 11,464/1846

Silencer (for guns) 1905
Hiram Percy Maxim (Britain)
Patent no: GB 17,799/1905

**Sliced bread (bread slicing and
wrapping machine) 1928**
Otto Frederick Rohwedder (USA)
Patent no: US 1,867,377
(filed 1928, granted 1932)

Slinky 1945
Richard T. James (USA)
Patent no: GB 630,702
(filed 1945) & US 2,415,012
(filed 1946, granted 1947)

Smokeless cartridges 1888
Hiram Stevens Maxim
(USA–Britain)
Patent no: GB 16,213/1888 &
US 430,212

Spider ladder 1994
Edward Thomas Patrick Doughney

(Britain)
Patent no: GB 2,272,154

**Square-bottomed grocery bag
1872**
Luther Childs Crowell (USA)
Patent no: US 123,811
(filed 1867, granted 1872)

**Square-bottomed grocery bag
(machinery for manufacture)
1872**
Luther Childs Crowell (USA)
Patent no: US 123,812
(filed 1867, granted 1872)

Steam-powered gun 1824
Jacob Perkins (USA)
Patent no: GB 4,952/1824

**Stripping film (paper-backed
photographic film) 1884**
George Eastman (USA)
Patent no: US 306,594

Submarine (first modern) 1898
John Philip Holland (Ireland–USA)
Patent no: US 708,553 (granted
1902)

**Substance dispensing headgear
1999**
Randall D. Flann (USA)
Patent no: US 5,966,743

Synthetic progesterone 1951
Carl Djerassi et al (Austria–USA)
Patent no: US 2,744,122

Synthetic progesterone 1953
Frank Colton (Poland–USA)
Patent no: US 2,691,028 &
US 2,725,389

Syringe (disposable) 1953
Henry G. Mollinari for Becton
Dickinson (USA)
Patent no: US 2,684,556
(filed 1953, granted 1954)

Tarmac 1902
Edgar Purnell Hooley (England)
Patent no: GB 7,796/1902 &
US 765,975

Tear-open can 1970
William Cookson (England)
Patent no: GB 1,213,616

Tear-strip drinks can 1963
Ermal Fraze (USA)
Patent no: US 3,255,917
(filed 1963, granted 1966)

**Teflon (polytetrafluoroethylene)
1938**
Roy J. Plunkett (USA)
Patent no: US 2,230,654
(filed 1939, granted 1949) & GB
625,348)

Telephone 1876
Alexander Graham Bell
(Scotland–USA)
Patent no: US 174,465 &
GB 4765/1876

Television 1923
John Logie Baird & Wilfred Day
(Scotland & England)
Patent no: GB 222,604

(granted 1924)

**Television (cathode ray
transmitter) 1923**
Vladimir Zworykin (Russia–USA)
Patent no: US 2,141,059
(granted 1938)

**Television (image dissector)
1927**
Philo T. Farnsworth (USA)
Patent no: US 1,773,980
(granted 1930)

**Tennis (portable court for lawn
tennis) 1873**
Walter Clopton Wingfield (Wales)
Patent no: GB 685/1874 (filed &
granted 1874)

Terylene 1941
John Rex Whinfield & James
Tennant Dickson (Britain)
Patent no: GB 578,079
(filed 1941, granted 1946)

Thermostat 1830
Andrew Ure (Scotland)
Patent no: GB 6,014/1830

Thermostat (bellows type) 1903
W.M. Fulton (Britain)
Patent no: GB 11,584/1903

Tiddlywinks 1888
Joseph Assheton Fincher
(England)
Patent no: GB 16,215/1888

Traffic signal 1922
Garrett Augustus Morgan (USA)
Patent no: US 1,475,024

Transistor (junction) 1947–50
William Shockley (USA)
Patent no: US 2,569,347
(granted 1950)

Transistor (point-contact) 1947
John Bardeen & Walter Brattain
(USA)
Patent no: US 2,524,035
(filed 1948, granted 1950) &
GB 694,021

Triumph swing stopper 1885
William Painter (USA)
Patent no: US 315,655

Tupperware 1947
Earl Silas Tupper (USA)
Patent no: US 2,487,400
(filed 1947, granted 1949) &
GB 662,219

Typewriter (concept) 1714
Henry Mill (England)
Patent no: GB 395

**Typewriter (commercially
manufactured) 1865**
Malling Hansen (Denmark)
Patent no: GB 1,385/1870

**Unicorns (method of growing)
1984**
Timothy G. Zell (USA)
Patent no: US 4,429,685

Vaseline (petroleum jelly) 1872
Robert A. Chesebrough

(England–USA)
Patent no: US 127,568 &
GB 1,012/1874

Velcro 1941
George de Mestral (Switzerland)
Patent no: GB 721,338 &
US 2,717,437 (filed 1951 in
Switzerland, GB patent granted
1955)

Viagra (sildenafil citrate) 1990
Andrew Simon Bell et al (Britain)
Patent no: EP 463,756 &
US 5,250,534 et al

Viagra (use for impotence) 1993
Nicholas Terrett & Peter Ellis
(Britain)
Patent no: WO 94/28902

**Voltage indicator (first patent in
the field of electronics) 1880**
Thomas Alva Edison (USA)
Patent no: US 307,031
(granted 1883)

**Vote-counting machine (Edison's
first patent) 1868**
Thomas Alva Edison (USA)
Patent no: US 90,646
(granted 1869)

**Walking on water (apparatus for)
1914**
Marout Yegwartian (England)
Patent no: GB 229/1914
(filed 1914, granted 1915)

**Wheeldex (precursor of Rolodex)
1940s**
Arnold Neustadter & Hildaur
Nielson (USA)
Patent no: US 2,598,819
(filed 1950, granted 1952)

Widget 1985
Alan James Forage & William
John Byrne for Arthur Guinness
Son & Co. (England & Ireland)
Patent no: GB 2,183,592 &
US 4,832,968 & EP 227,213

Wine box 1963
Thomas Angove (Australia)
Patent no: AUS 280,826
(granted 1965)

Workmate workbench 1967
Ron Hickman (South Africa)
Patent no: GB 1,267,032-5 &
US 3,615,087 (both filed 1968)

Zip 1913
Gideon Sundback (Sweden–USA)
Patent no: US 1,219,881
(filed 1914, granted 1917) &
GB 12,261/1915

**Zippity-Do ('Actuator for slide
fastener') 1971**
Marion Donovan (USA)
Patent no: US 3,599,274

INDEX

SELECTED BIBLIOGRAPHY

Anderson, Clive. *Patent Nonsense, A Catalogue of Inventions that Failed to Change the World.* Michael Joseph, 1994.

Baker, Ronald. *New and Improved: Inventors and Inventions that have Changed the World.* British Museum Publications Ltd, 1976.

Baren, Maurice. *How It All Began.* Smith Settle Ltd, 1992.

Brown, Travis. *Popular Patents.* Scarecrow Press Inc., 2000.

Giscard d'Estaing, Valérie-Anne. *The Second World Almanac Book of Inventions.* World Almanac, 1992.

Harris, Melvin. *ITN Book of Firsts.* Michael O'Mara Books, 1994.

Harrison, Ian. *The Book of Firsts.* Cassell Illustrated, 2003.

Harrison, Ian. *Action Man: The Official Dossier.* Collins, 2003.

Hillman, David & David Gibbs. *Century Makers.* Weidenfeld & Nicholson, 1998.

Johnson, Joel. "Jet Ski Evolution." Article in Boats.com, 2000.

Kennedy, Carol. *ICI: The Company That Changed Our Lives.* Paul Chapman Publishing, 1993.

Kim, Irene. "Handheld Calculators" (article in *Mechanical Engineering* magazine). ASME, 1990.

Moore, Pete. *E=mc²: The Great Ideas that Shaped Our World.* Friedman/Fairfax, 2002.

Parry, Melanie (ed). *Chambers Biographical Dictionary.* Chambers, 1999.

Petroski, Henry. *The Evolution of Useful Things.* Vintage Books, 1994.

Robertson, Patrick. *The Shell Book of Firsts.* Ebury Press & Michael Joseph, 1974.

Tibballs, Geoff. *The Guinness Book of Innovations.* Guinness Publishing, 1994.

Uhlig, Robert (ed). *James Dyson's History of Great Inventions.* The Telegraph Group, 2000.

van Dulken, Stephen. *Inventing the 20th Century.* The British Library, 2000.

van Dulken, Stephen. *Inventing the 19th Century.* The British Library, 2001.

ACKNOWLEDGMENTS

To CEA, with love

Thank you to the following people and organizations for their help in researching *The Book of Inventions*: Caroline Allen, Susie Allwork (Aerial Camera Systems), Clive Bacon (3M; www.mmm.com), David Baker (Land Shark Ltd; www.landshark.co.uk), Charles Brathwaite, Chris Bunting (L. Robinson & Co. [Gillingham] Ltd; www.jubileeclips.co.uk), Garth Conboy (Gemstar eBook Group Ltd), Jonathan Copus (Dentron Ltd), Gerd Cypers (Dolmar GmbH), James Donovan, Christine Donovan, Suzanne Durban (L. Robinson & Co. [Gillingham] Ltd), Hans Fairley (Andreas Stihl Ltd), Alex Friend (The Myelin Project), Art Fry, Cressida Granger (Mathmos), Mandy Haberman, Phil Harrison, Mary Harrison, Ulrikke Heggelund (Royal Norwegian Embassy, London), Ron Hickman, John Jennings (L. Robinson & Co. [Gillingham] Ltd), Ziaad Kahn (British Library Patent Department), Karl Kooiman (Rubbermaid [Rolodex]), Ruth Magnus (Institute of Patentees & Inventors), Celia Mannings, Heather McMahon (3M Corporate Communications), Ray Meads (3M), Peter Milner (Serraglaze), Kitty Minassian (Inventorlink; www.inventorlink.co.uk), Sue Newey (Trebor Bassett), Ingelise Nielsen (IDEO), Augusto Odone (The Myelin Project), Alison Oswald (Smithsonian Institution Archives Center), Richard Paine (Inventorlink), Johann Parkhill, Richard Penfold, Michelle Poole (Firefly Communications), Matt Redin (Angove's Wines), Mrs S. Rice (Marlow Foods Ltd [Quorn]), Chris Sachs, Barrie Singleton, John Singleton, Marie-Claire Walton, Barrie Wills (Serraglaze), John Wright (3M). Dayton Reliable Tool & Mfg Co., Dyson, ICI, Norwegian Science Museum, Philips Electronics, Tetra Pak, Wimbledon Lawn Tennis Museum.